The International Library of Bioethics

Volume 116

The *International Library of Bioethics* comprises international and interdisciplinary volumes on foundational and applied issues in bioethics. The series focuses on perennial topics of aging, mental health, preventive medicine, medical research issues, end of life, biolaw, and other areas of bioethics, whilst expanding into other current and future questions involving the living world. The *International Library of Bioethics* examines the role of bioethics against the background of increasing globalization and interdependency of the world's cultures and government.

The series explores meaningful ways to answer the questions of how to direct biotechnology to human and other living things' ends, how to deal with changed values in the areas of religion, society, and culture, and how to formulate a new way of thinking. That is, to create a new, practical bioethics.

The international Library of Bioethics welcomes book proposals representing the broad interest of the series' focus. We especially encourage proposals addressing aspects of changes in biological and medical research and clinical health care, health policy, medical and biotechnology, and other applied ethical areas involving living things, with an emphasis on those interventions and alterations that force us to re-examine foundational issues.

Kevin R. Smith
Editor

Bioethics and Religion

Exploring the Intersection

Springer

Editor
Kevin R. Smith
Faculty of Social and Applied Sciences
Abertay University
Dundee, UK

ISSN 2662-9186 ISSN 2662-9194 (electronic)
The International Library of Bioethics
ISBN 978-3-032-33537-1 ISBN 978-3-032-33538-8 (eBook)
https://doi.org/10.1007/978-3-032-33538-8

This work was supported by Wellcome Trust.

This Springer imprint is published by the registered company Springer Nature Switzerland AG
The registered company address is: Gewerbestrasse 11, 6330 Cham, Switzerland

Preface

This book is built around a central question: Should religion be permitted a seat at the table of bioethics? This inquiry lies at the heart of ongoing debates about the role of religious perspectives in addressing complex moral questions in medicine, healthcare, and biomedical science.

The aim of this volume is not to resolve this question definitively but to explore it from a wide range of perspectives. Contributors, recognized experts in their respective areas of ethical inquiry, have been invited to present their arguments on whether religion should play a role in shaping bioethical decision-making, focusing on specific topics within the field. These chapters include arguments from both secular and religious viewpoints, ensuring a balanced and intellectually stimulating debate.

The book is primarily concerned with high-level ethical and philosophical reasoning—that is, critically examining claims that religion has a valid role to play in normative bioethical discourse. At the same time, it also engages with issues of direct relevance to clinical practice and public policy. Some chapters explore the challenges faced by healthcare professionals navigating conscience-based refusals or provisions in pluralistic settings, while others consider how religious perspectives may be expressed or constrained within liberal policymaking frameworks. In this way, the volume offers insight into how foundational ethical reasoning connects with real-world professional and societal contexts.

Structurally, the volume is organized thematically. The general format is that for each topic, there is a chapter from a secular or atheistic perspective that critically examines the idea that religion offers something meaningful to bioethics and argues against its inclusion, and at least one chapter from a religious perspective—this may reflect a general theistic viewpoint or a specific tradition from a major world religion—defending the role of religion in bioethical discourse. This structure is intended to enable readers to compare contrasting arguments on the same issue, illuminating the diverse ways in which secular and religious perspectives approach bioethical challenges.

As the editor, I approach this project from a secular and atheistic perspective, while fully acknowledging the value of engaging with diverse viewpoints to enrich the discourse. By inviting contributions from prominent scholars representing both secular and religious perspectives, I have sought to capture the complexity and breadth of thought in this critical field. The resulting collection of compelling arguments aims to serve as a comprehensive resource for anyone interested in the ethical dimensions of modern medicine and science.

This volume is designed for a wide audience, including bioethicists, physicians, biomedical scientists, healthcare practitioners, and policymakers, offering a platform to explore the dynamic interplay between religion and secularism in bioethics. Regardless of a reader's personal stance, the book seeks to provoke thoughtful reflection and foster a deeper understanding of whether religion holds any validity in the domain of bioethics as a constructive force worthy of respect and consideration—or should instead be regarded as an inherently malign influence that ought not to be permitted a seat at the table of bioethics.

I wish to extend my sincere gratitude to all the contributing authors who have made this volume possible. The intersection of religion and bioethics is not only challenging but deeply controversial, often provoking intense debate and, at times, resistance to engagement. In this context, I am particularly appreciative of the authors' willingness to contribute their expertise to this scholarly dialogue, even when navigating sensitive and potentially divisive issues. Their thoughtful and rigorous contributions reflect an impressive depth of scholarship, and their openness to responding constructively to feedback has been invaluable in shaping this collection. Their dedication and intellectual generosity in addressing such a contentious topic are to be commended, and I am deeply grateful for the time and effort they have invested in this important collaborative endeavor.

Dundee, UK Kevin R. Smith

Contents

Chapter 1
Religion and Secularism in Bioethics: A Critical Introduction

Kevin R. Smith

Abstract This introductory chapter outlines and examines the contributions to this volume, exploring whether religious ethics enhance bioethical discourse or if secular frameworks provide a more universally applicable foundation. Themes such as autonomy, justice, and public reasoning are considered across major topics including abortion, conscience, medically assisted dying, circumcision, and enhancement. While religion offers cultural and moral depth, its doctrinal rigidity often clashes with the pluralistic and evidence-based demands of contemporary bioethics. By contrast, secular approaches, grounded in inclusivity and empirical validity, arguably emerge as more robust tools for addressing complex healthcare ethics. This chapter serves as both a thematic overview and critical evaluation, and seeks to set the stage for the deeper explorations of these tensions provided in the following chapters.

1.1 Preamble

Bioethics occupies a contentious space where deeply held beliefs about life, death, and human dignity frequently collide. Among these, few topics provoke as much debate as the role of religion in bioethical discourse. Should religion be granted a voice in public bioethical discourse, or does its reliance on faith and doctrine undermine the universality and inclusivity demanded of such debates? This chapter is intended to serve as a critical entry point, setting the stage for the contributions to follow. Here, I outline the chapters and themes, and invite readers to reflect on the interplay between secular and religious perspectives, not merely as abstract philosophical positions but as forces that shape real-world decisions in medicine, healthcare, and science. Whether religion enriches or detracts from bioethics is a question

K. R. Smith (✉)
Faculty of Social and Applied Sciences, Abertay University, Dundee, UK
e-mail: k.smith@abertay.ac.uk

K. R. Smith (ed.), *Bioethics and Religion*, The International Library of Bioethics 116, https://doi.org/10.1007/978-3-032-33538-8_1

that will resurface throughout this volume, one that demands rigorous analysis and, perhaps, a measured dose of intellectual humility.

This chapter is structured to introduce readers to the authors' competing positions on the topics they have engaged with. For each topic, I begin by offering a very brief overview of the competing positions presented by the authors. Following this, I set out a short summary of each author's arguments in turn. Next, I examine the key themes and points of divergence that emerge from these contributions, identifying the central tensions and areas of overlap. Finally, I provide my own commentary, where I strive to offer a balanced and critical reflection on the strengths and weaknesses of the arguments presented. This structure is designed to facilitate engagement with the various perspectives explored in this volume, and I hope it will effectively set the stage for deeper examination in the chapters that follow. I acknowledge from the outset that I am a secular bioethicist, and my interpretations are inevitably influenced by that perspective. And needless to say, I am responsible for any errors of fact or interpretation regarding the contributing authors' positions.

1.2 Jonathan Baron's "Rational Bioethics Without Religion" and Michael McCarthy's "Towards a Pluralistic and Public Bioethics: Interreligious Dialogue as a Paradigm for Understanding"

Baron and McCarthy present contrasting views on the role of religion in the overall enterprise of bioethics. Baron champions a secular, utilitarian framework grounded in actively open-minded thinking (AOT), while McCarthy argues for the inclusion of religious perspectives through a pluralistic and dialogical approach.

1.2.1 Baron: Rationality and Active Open-Minded Thinking

Baron situates his argument within a framework of utilitarianism, proposing that bioethics should prioritize benefit vs. harm, fairness, and the rejection of pseudo-facts derived from religious doctrine. He critiques the influence of religion on moral decision-making, particularly its reliance on tradition and authority. Baron's concept of actively open-minded thinking (AOT) aims to foster impartiality and reduce cognitive biases, and thus promote a form of moral reasoning unencumbered by religious dogma.

Central to Baron's thesis is the principle of universality: moral reasoning should employ tools such as the "veil of ignorance" and the "ideal observer" to prioritize collective welfare. He argues that these methodologies expose the inconsistencies and parochialism inherent in religiously motivated norms, such as the Doctrine of Double Effect in Catholic theology. By adopting AOT, Baron contends, individuals and institutions can better align ethical judgments with rational and evidence-based principles. Baron's approach offers a rigorous framework for critical thinking and is likely to resonate with utilitarians and others who prioritize logic and reason as foundational principles in bioethics.

1.2.2 McCarthy: Interreligious Dialogue and Pluralistic Bioethics

In contrast to Baron's reliance on utilitarian reasoning, McCarthy advocates for a pluralistic approach, emphasizing the importance of interreligious dialogue as a means of enriching bioethical discourse. Drawing on Charles Taylor's concept of the "secular age" and the reforms of the Second Vatican Council, McCarthy highlights the potential for religious and secular perspectives to engage constructively. He frames bioethics as an interdisciplinary practice that should transcend sectarian divides, fostering mutual understanding and respect.

McCarthy's vision incorporates theological insights into broader public discussions, using examples from Catholic Social Teaching to address issues such as healthcare inequities and social determinants of health. He emphasizes that interreligious dialogue does not seek to convert or dominate but rather to illuminate the moral reasoning behind diverse perspectives. By adopting this style of discourse, McCarthy argues, bioethics can address complex ethical challenges while respecting cultural and religious diversity.

1.2.3 Key Themes and Points of Divergence

The contrasting approaches of Baron and McCarthy highlight fundamental differences in ethical reasoning. Baron champions a secular framework, emphasising actively open-minded thinking (AOT) and utilitarian principles as universal tools for bioethics. In contrast, McCarthy seeks to integrate religious insights into public bioethics, proposing interreligious dialogue as a bridge to connect diverse cultural and ideological perspectives. This divergence reflects deeper tensions about the role of religion in ethical discourse. Baron views religion as a source of cognitive bias and pseudo-facts, arguing for its exclusion from public bioethical deliberations. McCarthy, however, argues that religious traditions bring valuable perspectives to moral reasoning, particularly through their historical and cultural contributions. These differing approaches also shape their views on inclusivity and practicality. While Baron's framework prioritises impartiality and universality, it risks alienating religious communities by sidelining their beliefs. McCarthy's pluralistic model, by contrast, fosters inclusivity but faces potential challenges in achieving consensus on deeply contentious issues. Together, their perspectives illustrate the ongoing tension between universal principles and the contextual aspects of cultural and religious diversity in bioethical practice.

1.2.4 Editor's Commentary

While McCarthy envisions a pluralistic framework where religious and secular perspectives engage as equals, Baron argues that bioethics must be rooted in rational, evidence-based reasoning to ensure intellectual rigor and universal applicability. McCarthy's reliance on interreligious dialogue as a model for public bioethics assumes parity between secular and religious frameworks. Baron's analysis challenges this assumption by highlighting the fundamental epistemic divide between the two. Secular bioethics, as Baron argues, is grounded in principles like benefit/harm and fairness, which are subject to rational scrutiny and accessible to all rational agents. Religious bioethics, by contrast, often derives its legitimacy from doctrinal authority or sacred texts, which lack universal acceptance and are not open to critical examination. I suggest that unless religious bioethicists can provide justification for why religious perspectives should be treated as intellectual equals to secular reasoning, the universality and objectivity criteria that public bioethics requires will remain unmet.

Baron's concept of actively open-minded thinking (AOT) further underscores the limitations of religious ideas in public discourse. AOT emphasizes the importance of questioning one's assumptions, seeking counterarguments, and avoiding confirmation bias. Religious frameworks, as Baron notes, are often resistant to such scrutiny because they are grounded in immutable doctrines rather than adaptable, evidence-based reasoning. It appears to me that McCarthy's assertion that interreligious dialogue fosters mutual understanding does not resolve this epistemic gap; instead, it arguably risks conflating doctrinal faith with rational deliberation, thus threatening to weaken the integrity of bioethical debate.

Moreover, Baron's critique of pseudo-facts—claims presented as moral absolutes without empirical justification—casts doubt on the viability of religious contributions to public bioethics. Religious arguments, such as those against abortion or euthanasia, often rely on unprovable assertions (e.g., the sanctity of life as defined by divine command) that cannot be evaluated within the framework of public reason. While McCarthy acknowledges the challenges of reconciling religious and secular perspectives, he arguably does not adequately address how religious ideas, which often fail to meet the standards of rational justification, can meaningfully contribute to public bioethical discourse.

Baron's framework also highlights the asymmetry in the acceptance of secular versus religious ideas. Secular principles, such as the utilitarian emphasis on maximizing well-being, are rooted in reasoning that any rational person can, in principle, accept. Religious principles, however, are frequently tied to specific metaphysical beliefs that are not universally shared. McCarthy's call for interreligious dialogue does not, in my view, persuasively account for this fundamental disparity. As Baron points out, even when religious arguments are "translated" into secular terms, this process risks bad faith and intellectual disingenuity, as it obscures the theological foundations of the original argument.

In conclusion, while McCarthy's chapter seeks to promote dialogue and inclusivity, Baron's insights reveal the inherent challenges of integrating religious perspectives into public bioethics. The former's assumption of parity between religious and secular frameworks would require justification, given the epistemic and methodological advantages of the latter. Religious ideas often lack the transparency and rational basis required for participation in public reason, and their inclusion risks diluting the intellectual rigor of bioethical discourse. By contrast, a bioethics grounded in rational, actively open-minded thinking appears to provide a more robust and universal foundation for making progress in bioethics.

1.3 "Conscience Beyond Religion: Exploring Secular Conscientious Objection and Provision" by Abram Brummett, "Conscientious Objection in Healthcare" by Xavier Symons, and "Conscientious Objection in Healthcare: A Jewish Ethical Perspective on Balancing Professional Integrity and Patient Care" by Jason Weiner

The role of conscience in healthcare bioethics represents a critical fault line in debates over moral integrity, autonomy, and professional responsibility. Brummett, Symons, and Weiner offer contrasting perspectives on this theme, engaging with the concept of conscientious objection from secular, philosophical, and Jewish ethical lenses, respectively. Their chapters explore the tension between respecting individual conscience and ensuring equitable patient care.

1.3.1 Conscience and Secular Ethics: Brummett's Argument

Brummett advocates for the inclusion of secular conscientious objection alongside traditional religious objections. The conscience debate has predominantly focused on safeguarding the rights of clinicians who, on religious grounds, seek to refuse participation in legal and professionally accepted medical treatments. In response, some secular bioethicists have called for the wholesale rejection of conscience claims in healthcare. However, such arguments have largely failed to effect legal reform, and, conversely, conscience rights have only expanded over time. In his chapter, Brummett adopts a different approach by examining the potential for secular conscientious objection and provision. He argues that if conscience is to remain an inescapable aspect of healthcare, its applications beyond religious contexts warrant thorough and sustained exploration.

Brummett frames conscience as a publicly defensible moral construct, applying the "reasonability view," which permits only those claims of conscience that can withstand scrutiny under public reason. His arguments extend to situations where clinicians refuse to treat neurologically devastated patients or seek to provide services prohibited by institutional policies, such as those in Catholic hospitals. Brummett highlights the asymmetry in legal protections, arguing that conscientious provisions (e.g., referrals) deserve similar protections as objections, provided they impose minimal institutional burdens.

1.3.2 The Virtue of Moral Conscience: Symons' Philosophical Defense

ymons constructs a philosophical defense of conscientious objection, rooted in the dual foundations of conscience as a tracker of moral truth and an expression of moral integrity. Symons situates conscientious objection within liberal democratic principles, emphasizing the importance of accommodating moral and ethical disagreements in pluralistic societies. He contrasts conscientious objection with civil disobedience, arguing that the former seeks to preserve individual moral integrity without undermining the law.

1.3.3 Jewish Ethical Perspectives: Weiner's Pragmatic Framework

Weiner approaches conscientious objection through a Jewish ethical lens, emphasizing the principle of "placing a stumbling block before the blind" as a metaphor for assessing moral complicity. This results-based approach permits participation in morally contentious actions when the act would proceed regardless of one's involvement, thereby minimizing direct moral compromise. Weiner's framework allows for qualified participation, such as providing referrals or ancillary support, and underscores the importance of balancing professional integrity with patient care.

1.3.4 Key Themes and Tensions

Brummett and Symons take distinct positions on the universality of conscience. Brummett advocates for a secular framework that seeks publicly defensible standards for conscience, while Symons emphasises a theological orientation, grounding conscience in objective morality. Weiner, offering a Jewish perspective, reframes conscience as inherently context-dependent, viewing it through the lens of communal and relational ethics.

All three authors grapple with the delicate balance between respecting clinician integrity and ensuring patient rights. Brummett focuses on pragmatic solutions that minimise institutional burdens, particularly for conscientious provisions, and Symons defends the presumption of respect for conscientious objection while recognising the need for limitations to prevent harm. Weiner's application of Jewish legal principles advocates for approaches that uphold professional integrity while safeguarding patient access to care.

The implications for policy further underscore these tensions. Brummett and Symons highlight the need to protect conscience rights in pluralistic societies, with Brummett calling for equitable consideration of both objections and provisions. Meanwhile, Weiner suggests practical accommodations that align with both religious and ethical pluralism.

1.3.5 Editor's Commentary

The conscience debate has traditionally been framed by religious bioethicists, theologians, and practitioners as a means to protect clinicians who refuse to provide legal and professionally accepted medical treatments on the basis of religious beliefs. In contrast, secular bioethicists have predominantly called for the outright rejection of conscience claims in healthcare. Brummett, Symons, and Weiner, however, present intriguing perspectives that challenge this binary framing.

Brummett advances a secular defense of conscience that is fully compatible with public reason, arguing that healthcare professionals should have the right to conscientiously refuse to provide treatment, particularly in cases involving neurologically devastated patients. His argument is not based upon religious justifications typically associated with conscientious objection, instead positioning conscience as a universal moral faculty that can coexist within a pluralistic healthcare system.

Symons also defends conscientious objection, and extends his reasoning to both individual and institutional contexts. His approach is distinct from Brummett's, asserting that individual objection deserves respect because it reflects a person's sincere effort to discern moral truth. Symons also highlights the importance of institutional conscience, arguing that organizations require a degree of autonomy to act in accordance with their mission and values, framing such autonomy as a societal good. His dual focus on individual and institutional conscience is an unusual and valuable contribution. However, institutional conscience could conflict with the ethical principle of justice by creating access barriers for patients. For instance, a Catholic hospital's refusal to provide contraception might leave patients in underserved areas without alternatives. And Symons' argument that conscience reflects an agent's "best effort to get at moral truth" invites debate about the subjective versus objective grounding of moral truths.

Weiner approaches the issue from a distinctively Jewish perspective, offering an ethical analysis rooted in Jewish law. He critiques the assumptions of secular bioethicists who often dismiss religious perspectives as irrational or overly restrictive. While many religious viewpoints uphold conscientious objection as a moral imperative, Weiner argues that Jewish principles may, in some cases, lead to the opposite conclusion. He contends that conscientious objection should not always be the default response, particularly when it risks causing greater harm or undermining the broader goals of care. Instead, he posits that moral integrity can often be preserved without invoking conscientious objection, emphasizing the need to balance personal values with the practical realities of healthcare. Weiner's use of the Jewish ethical framework introduces a pragmatic, results-oriented approach, shifting the focus from the objector's moral integrity to the broader outcomes of their actions. However, his arguments risk oversimplifying the complex interplay between personal integrity and professional obligations. For instance, his leniency in indirect involvement might not resonate with professionals who feel morally compromised even by tangential participation. And his "same side of the river" paradigm, while innovative, is difficult to reconcile with secular frameworks, which focus on the direct versus indirect causation of harm.

Collectively, these three authors offer what amounts to a unique and balanced view of conscientious objection, moving beyond the polarized positions that dominate the debate. Their perspectives challenge the typical strong opposition of secular bioethicists to conscientious objection and the uncritical support often voiced by religious proponents. Their arguments also expose unresolved tensions—between individual and institutional claims, between public reason and personal belief, and between moral integrity and systemic justice. By addressing these tensions, the chapters provide unique and valuable insights into how contemporary bioethics can navigate the complex and practical ethical challenges posed by conscientious objection.

1.4 Alberto Giubilini's "Should Religion Be Permitted a Seat at the Table of Bioethics? The Case of Abortion" and Robert P. George & Christopher Tollefsen's "Abortion, Religion, and Bioethics"

The topic of abortion remains one of the most contentious areas in bioethics, where moral, religious, and philosophical perspectives intersect. Giubilini and George & Tollefsen provide opposing frameworks for the role of religion in bioethical debates on abortion. Giubilini challenges the sufficiency and necessity of religious reasoning in liberal societies while acknowledging its instrumental value. In contrast, George & Tollefsen emphasize the compatibility of Catholic natural law reasoning with public discourse and defend religious participation in bioethical discussions.

1.4.1 Giubilini: The Instrumental Value of Religion in Liberal Bioethics

Giubilini begins with a general exploration of whether religion should participate in bioethical discussions, distinguishing between bioethics, healthcare professionalism, and public policy. While conceding that religious perspectives can contribute to the moral debate on abortion, he argues that religious values must undergo public scrutiny to ensure their compatibility with liberal principles of inclusivity and reason.

Giubilini cites natural law theory to highlight overlaps between religious and secular views, such as arguments against abortion based on teleological understandings of human life. He observes that natural law theory does not inherently require religion; instead, it operates through reason, which is universally accessible. Consequently, religion's role in abortion debates becomes instrumental rather than foundational, with faith and authority-based arguments relegated to the private sphere in liberal societies.

Giubilini also critiques the use of Christian notions of equality to oppose abortion. While acknowledging their historical influence on liberal democracy, he argues that such principles are not universally applicable without additional justification. Non-religious perspectives, he suggests, can reasonably exclude fetuses from moral equality based on a lack of relevant capacities, such as self-awareness or rationality.

1.4.2 George & Tollefsen: Natural Law and the Public Role of Religion

George & Tollefsen adopt a robust defense of Catholic natural law reasoning in bioethics, framing abortion as a violation of fundamental human dignity and the common good. Their arguments rest on four propositions: that human life begins at conception, that all humans share equal moral worth, that the state has a duty to protect prenatal life, and that such protections are defensible without appeal to religious revelation.

Unlike Giubilini, George & Tollefsen do not regard natural law reasoning as a secular framework distinct from religion but as a bridge between faith and reason. They argue that natural law ethics aligns with Catholic teaching while also providing a rational foundation for public debate. For them, excluding religious voices from abortion discussions undermines the principles of equality and civility in liberal democracies.

They critique Rawlsian public reason for sidelining religious arguments, highlighting the civil rights movement as an example of effective faith-based advocacy. However, they acknowledge the limitations of purely sectarian approaches and advocate for translating religious convictions into terms accessible to a pluralistic audience.

1.4.3 Key Points of Divergence

The points of divergence between Giubilini and George & Tollefsen reveal fundamental tensions in the application of natural law, the role of religion in public policy, and the moral status of the fetus. Giubilini conceptualises natural law as a secular construct, albeit one that is compatible with and indeed utilized within Christianity, emphasising its inherent independence from faith-based premises and its capacity to transcend religious frameworks. In contrast, George & Tollefsen integrate natural law with Catholic doctrine, presenting it as a shared moral grammar that bridges the perspectives of believers and non-believers alike.

The authors also differ in their views on religion's role in public policy. Giubilini is cautious, arguing that appeals to faith or religious authority undermine the liberal principles of inclusivity and universality. He advocates for religious participation in public debates only when its arguments are translated into terms that are universally accessible. George & Tollefsen, however, reject such restrictions, asserting that even explicitly sectarian arguments have a rightful place in public discourse, provided they engage respectfully and constructively with opposing views.

On the moral status of the fetus, both perspectives acknowledge its possible significance but diverge in their justificatory frameworks. Giubilini prioritises rational scrutiny of natural law claims, allowing for secular disagreement over the fetus's moral status. George & Tollefsen, by contrast, ground their argument in the assertion that human dignity is intrinsic, based on capacities present from conception. They view this position as both rationally defensible and consistent with Catholic teaching.

1.4.4 Editor's Commentary

Giubilini and George & Tollefsen offer sharply contrasting perspectives on the relationship between religion and bioethics, particularly in the contentious domain of abortion. Both address the interaction of religious values and public reason, yet their treatments diverge significantly in their approach and implications.

Giubilini's reliance on public reason aligns with secular liberal principles, proposing that ethical arguments must be universally accessible to participants in pluralistic societies. He takes trouble to avoid reliance on a 'straw man' pastiche of the religious viewpoint. However, he strongly critiques the sufficiency and necessity of religion in bioethical discussions, arguing that its contributions, even if valuable from an instrumental or procedural perspective, must be disentangled from appeals to faith or divine authority. He emphasizes that natural law, which is central to Christian opposition to abortion, can be used to generate arguments that are intelligible and defensible without invoking religious doctrine.

In contrast, George & Tollefsen defend a natural law perspective rooted in Catholic moral tradition but claim its principles are accessible through reason alone.

They advocate for the full participation of religious voices in public debates, cautioning against the exclusion of sectarian views under the guise of Rawlsian public reason. Their insistence on the universal applicability of their natural law framework, derived from a distinctly Catholic moral anthropology, risks undermining inclusivity by privileging a conception of human dignity that appears incompatible with secular perspectives such as that of Giubilini.

Both chapters address the challenge of integrating religiously grounded ethical views into public bioethics while maintaining pluralism. Giubilini navigates this terrain with a rigorous commitment to secularism, avoiding a simplistic rejection of religion but emphasizing that contributions to bioethics must be framed in terms of universally accessible, publicly defensible reasons. George & Tollefsen, meanwhile, advocate for a natural law framework that they claim is universally rational, allowing religious traditions to present their perspectives as truths derived from natural reason.

While this approach does not outright oppose Giubilini's insistence on public reason as a prerequisite for discussions on abortion and other bioethical issues, their framework diverges significantly in substance. George & Tollefsen assert that natural law renders all human life inherently equal, extending this principle to include embryos and fetuses. In contrast, Giubilini argues that religiously inspired claims to equality should be confined to a procedural recognition of equal participation in debate. For him, substantive issues—such as the permissibility of abortion—must be adjudicated based solely on arguments that meet the standards of public reason, free from theological underpinnings or sectarian assumptions.

1.5 Udo Schuklenk's "Secular Ethics Approaches to Medically Assisted Dying" and Mark J. Cherry's "Killing Off the Elderly, Sick, and Other Expensive Patients: Religious Voices in Public Bioethics Debate"

The debate on medically assisted dying (MAiD) offers profound insights into the intersection of autonomy, moral pluralism, and the role of religion in public bioethics. Schuklenk and Cherry provide sharply contrasting perspectives on this contentious issue, articulating secular and Christian frameworks, respectively. Schuklenk advocates for a public-reason-based justification of MAiD, while Cherry critiques its moral foundation, emphasizing the spiritual and societal dangers inherent in normalizing euthanasia and assisted suicide.

1.5.1 Schuklenk: Autonomy and Public Reason

Schuklenk grounds his support for MAiD in secular ethics, emphasizing autonomy, rationality, and public reason. He asserts that in liberal democracies, the separation of religion and state necessitates ethical frameworks accessible to all, regardless of faith. According to Schuklenk, MAiD aligns with principles of individual liberty, enabling decisionally capable individuals to determine their own fate, free from the constraints of religious doctrine.

Schuklenk identifies key criteria for MAiD: voluntariness, decisional capacity, stability of the request, and a clear understanding of one's situation. He defends these principles as safeguards against abuse and as consistent with liberal commitments to personal autonomy. He critiques religious opposition to MAiD, arguing that it often disguises theological imperatives under secular language, thereby undermining genuine public reason-based debate. By contrast, Schuklenk contends that secular ethics allows for a pluralistic and inclusive discourse that respects the diverse moral convictions of a liberal society.

While Schuklenk acknowledges concerns about slippery slopes and vulnerability, he maintains that these challenges can be mitigated through rigorous regulation. His framework champions individual choice and rejects paternalistic interventions, framing MAiD as a legitimate exercise of personal autonomy rather than a societal judgment on the value of life.

1.5.2 Cherry: Spiritual and Moral Objections

Cherry presents a robust critique of MAiD from a Christian perspective, framing euthanasia and assisted suicide as moral transgressions that violate divine sovereignty and human dignity. He argues that life and death fall under God's dominion, and that MAiD represents an affront to this natural order. Cherry extends this critique to secular justifications for MAiD, contending that without a canonical moral foundation, secular ethics devolves into moral relativism, incapable of resolving deep ethical disputes.

Cherry highlights the societal risks associated with MAiD, particularly the erosion of medical professionalism and the devaluation of vulnerable lives. He cites examples from jurisdictions such as Canada and the Netherlands, where eligibility for MAiD has expanded beyond terminal illness to encompass psychological suffering and non-terminal conditions. For Cherry, such developments exemplify the slippery slope of normalizing euthanasia, leading to a commodification of life and a loss of moral restraint.

Cherry's critique of autonomy-based arguments emphasizes the relational and social dimensions of human existence. He warns that framing MAiD as a personal choice neglects the societal pressures and economic factors that may influence individuals' decisions, particularly among the elderly and disabled.

1.5.3 Key Themes and Points of Divergence

The contrasting perspectives of Schuklenk and Cherry reveal significant divergences in their treatment of moral foundations, the role of autonomy, issues of vulnerability and social justice, and the contributions of religion to bioethics. Schuklenk's secular approach is grounded in autonomy and public reason as universal principles, explicitly rejecting the authority of religious doctrines in shaping public policy. In contrast, Cherry anchors his argument in Christian theology, asserting that divine sovereignty and moral absolutes form the foundation of ethical decision-making. This fundamental difference underscores the broader tension between secular and religious bioethics in establishing a shared moral framework.

Their views on autonomy further illustrate this divide. Schuklenk elevates autonomy as the cornerstone of ethical decision-making, advocating for the right of individuals to make self-regarding choices about their own lives and deaths. Cherry, however, challenges this perspective, highlighting the interconnected nature of human existence and the social responsibilities that limit purely individualistic conceptions of autonomy. For Cherry, autonomy must be tempered by an ethical framework that prioritises communal well-being and relational interdependence.

The issue of vulnerability and social justice further complicates their positions. Schuklenk acknowledges the risks of coercion and exploitation inherent in medically assisted dying (MAiD) but frames these as regulatory challenges that can be mitigated through proper oversight. Cherry, on the other hand, warns that MAiD perpetuates systemic inequalities, disproportionately impacting marginalised groups such as the disabled and elderly. He views this practice as reinforcing societal biases and devaluing the lives of those most vulnerable to external pressures.

Finally, the role of religion in bioethics emerges as a point of contention. Schuklenk critiques religious opposition to MAiD, arguing that theological arguments lack the universality required in pluralistic societies. For him, religious perspectives fail to meet the standards of public reason, making them ill-suited for shaping policy in a diverse and inclusive world. Cherry defends the inclusion of religious views, asserting that they provide moral clarity and depth often absent in secular frameworks. He sees religious ethics as a necessary counterbalance to the perceived moral relativism of secular bioethics. Together, these contrasting views highlight the enduring challenges of integrating religious and secular perspectives in addressing complex bioethical issues like MAiD.

1.5.4 Editor's Commentary

Schuklenk's chapter presents a philosophically rigorous defense of public reason as the foundation for bioethical deliberation. By centering autonomy and rational choice, Schuklenk highlights the importance of personal agency and individual well-being in ethical decision-making, offering a compelling case for the decriminalization of assisted dying.

One of the chapter's key strengths lies in its clear articulation of why religiously derived arguments should not shape public policy in liberal democracies. Schuklenk deftly critiques the covert use of theological concepts like "human dignity" and "vulnerability," exposing their lack of definitional clarity and their function as rhetorical tools to mask religious ideology. His analysis demonstrates how these terms, while superficially appealing, fail to meet the standards of public reason, thereby rendering them unsuitable for policymaking in pluralistic societies.

Schuklenk robustly defends the permissibility of assisted dying through both liberal and consequentialist frameworks. He effectively counters concerns about slippery slopes and coercion by emphasizing the safeguards inherent in well-regulated systems. His argument that respect for self-regarding choices is foundational to a good society is particularly persuasive.

While Schuklenk acknowledges the diversity of secular ethical perspectives, his commitment to public reason ensures that his analysis remains inclusive and accessible. This chapter serves as a model of how secular ethics can navigate complex bioethical issues without resorting to the exclusionary frameworks of religious dogma.

By contrast, Cherry's chapter advances an unabashedly religious defense of bioethics. Cherry is unflinching in his appeal to divine authority, prioritizing theological fidelity over broader inclusivity, a stance that is rare in contemporary bioethical discourse. In so doing, he in effect repudiates the thrust of Schuklenk's chapter. However, this same boldness is arguably the chapter's greatest weakness, as it inevitably alienates non-Christian audiences and thus undermines its relevance in pluralistic societies.

Cherry's reliance on theological premises, such as God's dominion over life and death, places his argument outside the domain of public reason, where moral claims must be accessible to all, regardless of religious belief. His insistence that euthanasia and assisted suicide are comprehensible only through divine revelation excludes secular interlocutors and fails to engage with widely accepted principles of autonomy and compassion.

The chapter also mischaracterizes secular ethics, portraying it as a relativistic "do-it-yourself project" devoid of definitive moral guidance. This caricature ignores established secular frameworks such as Kantian ethics, human rights theories, and utilitarianism, which offer robust justifications for moral action.

Despite these criticisms, Cherry's unapologetic commitment to a theological vision of bioethics is intellectually honest and refreshingly unguarded. Readers will ultimately need to decide for themselves whether such a resolutely religious approach can find a meaningful place in contemporary bioethical debates, particularly in liberal, pluralistic societies.

Together, these two chapters represent the stark divide between religiously grounded and secular approaches to bioethics. Schuklenk offers a compelling vision of bioethical deliberation rooted in inclusivity, rationality, and public reason, demonstrating how secular ethics can address complex issues like assisted dying without reliance on theological frameworks. In contrast, Cherry's chapter provides a provocative yet polarizing religious perspective that underscores the tensions between faith-based arguments and the demands of pluralistic, democratic discourse. Taken together, the chapters illuminate the critical importance of public reason in ensuring that bioethical debates remain accessible, equitable, and relevant to all members of society, regardless of their worldview.

1.6 Max Buckler & Brian D. Earp's "Circumcision as a Critical Test Case for the Value of Religious Bioethics: A Close Examination of the Writings of Rabbi Elliot N. Dorff" and Hossein Dabbagh's "Contested Cuttings: The Ethics of Genital Cutting in Islam Through the Lens of Peer Disagreement and Secular Bioethics"

The ethics of circumcision stands as a highly contested issue in bioethics, where debates over bodily autonomy, cultural identity, and religious freedom intersect. Buckler & Earp critique religious justifications for male circumcision, using Rabbi Elliot Dorff's Jewish framework as a lens to explore broader tensions within religious bioethics. Dabbagh offers an Islamic perspective, using the concept of peer disagreement to argue that certain genital cutting practices may be non-obligatory, even within religious contexts. Together, these chapters illuminate the complex interplay between religious traditions and secular bioethics in debates on non-therapeutic circumcision.

1.6.1 Buckler & Earp: The Inconsistencies of Religious Bioethics

Buckler & Earp's analysis interrogates the coherence of religious bioethics by examining Rabbi Elliot Dorff's defense of male circumcision. Dorff invokes the principle that "God owns the body" (GOTB) to justify circumcision as a sacred act within Judaism, symbolizing an immutable covenant with God. However, Buckler & Earp argue that this principle leads to contradictions. While GOTB ostensibly opposes non-voluntary body modifications, Dorff makes a significant exception for circumcision, relying on its religious significance to justify its ethical permissibility.

The authors further critique Dorff's inconsistent application of bodily integrity, contrasting circumcision with other body modifications—such as tattoos or genital piercings—which he opposes on theological and ethical grounds. Buckler &

Earp highlight the selective nature of Dorff's reasoning, which fails to apply GOTB consistently to similar practices. This inconsistency, they argue, undermines the validity of religious bioethical frameworks that privilege specific rituals over broader principles like bodily autonomy.

Additionally, Buckler & Earp examine Dorff's approach to non-Jewish genital cutting practices, such as Islamic ritual pricking. They question whether his framework can accommodate such practices without violating principles of religious pluralism. The authors conclude that Dorff's position exemplifies the broader challenges facing religious bioethics, which must reconcile doctrinal commitments with universal ethical principles.

1.6.2 Dabbagh: Peer Disagreement and Reflective Equilibrium

Dabbagh's chapter approaches circumcision through the lens of Islamic jurisprudence and peer disagreement, challenging the necessity of the practice within Islam. While male circumcision is widely practiced among Muslims, Dabbagh highlights significant variation in its religious interpretation. Some Islamic scholars view circumcision as obligatory, while others consider it recommended but non-essential. This epistemic uncertainty, he argues, opens the door for devout Muslims to critically reassess the practice.

Dabbagh employs the concept of reflective equilibrium to explore how religious adherents might balance conflicting moral intuitions and principles. He suggests that the absence of explicit Qur'anic endorsement for circumcision weakens its theological necessity, particularly when juxtaposed with ethical principles like nonmaleficence and bodily autonomy. Dabbagh's analysis extends to female genital cutting (FGC), noting that while it is less widely accepted in Islamic jurisprudence, the ethical challenges it poses are analogous to those of male circumcision.

Importantly, Dabbagh engages with secular bioethics to underscore the tensions between religious practices and modern principles of autonomy and harm avoidance. He contends that Muslim parents and communities could justify deferring circumcision until the individual can provide informed consent, aligning with both Islamic principles and secular ethical norms. This approach, he argues, respects the pluralistic nature of Islamic jurisprudence while prioritizing individual rights.

1.6.3 Key Themes and Points of Divergence

The ethical debates surrounding circumcision expose significant tensions in the treatment of bodily integrity, religious pluralism, and the intersection of religious and secular ethics. Buckler & Earp critique the inconsistent application of bodily integrity within religious bioethics, particularly in Dorff's defence of circumcision, which selectively permits non-voluntary bodily modifications based on religious tradition. In contrast, Dabbagh elevates autonomy as a universal ethical principle, arguing that the practice of circumcision should be deferred until an individual can provide informed consent. This emphasis on autonomy challenges the presumption that religious obligations should override the right to bodily self-determination.

The authors also diverge on the question of religious pluralism. Buckler & Earp express scepticism about the ability of religious bioethics to accommodate diverse traditions, questioning whether principles like "God owns the body" (GOTB) can be consistently applied across different faiths without leading to contradictions. Dabbagh, however, highlights the internal diversity within Islamic jurisprudence, using the concept of peer disagreement to argue for a pluralistic approach. He suggests that such pluralism allows for a critical reassessment of religious practices, enabling communities to adapt to contemporary ethical standards while respecting their traditions.

The intersection of religious and secular ethics further illuminates these debates. Buckler & Earp emphasise the limitations of religious frameworks in addressing secular concerns such as autonomy and harm, pointing to the need for ethical principles that transcend doctrinal boundaries. Dabbagh, by contrast, demonstrates how Islamic and secular bioethics can be integrated through reflective equilibrium, a process that seeks to balance moral intuitions with rational principles. This approach seeks to bridge the gap between tradition and modernity, offering a pathway for reconciling religious values with universal ethical norms.

1.6.4 Editor's Commentary

In these chapters, Buckler & Earp and Dabbagh explore the issue of the role of religion in bioethics by engaging with the fraught ethical terrain of circumcision. In so doing, they offer contrasting perspectives on the interplay between religious bioethics and secular principles. Buckler & Earp's critique of Rabbi Dorff's "God owns the body" (GOTB) framework is particularly incisive, exposing inconsistencies in applying this principle to justify male circumcision while opposing other bodily modifications. Their argument underscores a significant limitation of religious bioethical frameworks: their tendency to prioritize doctrinal commitments over universal principles like bodily autonomy and harm avoidance. Although their focus is on Judaism, the arguments they advance have broader implications that pertain to religious perspectives on circumcision in general.

Dabbagh, by contrast, utilises the internal pluralism of Islamic jurisprudence to argue for a reflective equilibrium between tradition and contemporary ethical standards. While his call to defer circumcision until informed consent can be given aligns with secular principles, and potentially applies not only to Islam but to religious perspectives in general, it arguably lacks the rigor of Buckler & Earp's broader critique, which incisively questions the fundamental coherence of religious bioethics in the context of circumcision.

Ultimately, Buckler & Earp's position is compelling in its universal applicability. Their focus on autonomy and pluralism highlights the ethical challenges of privileging specific religious practices over consistent moral principles. While Dabbagh's appeal to pluralism within Islam is noteworthy, Buckler & Earp offer a particularly strong foundation for navigating the intersection of religion and secular bioethics.

1.7 James J. Delaney's "Enhancement, Natural Law, and New Basic Human Goods"

Human enhancement has long sparked debates in bioethics, with critics highlighting its potential to alter human nature and proponents emphasizing its capacity to expand human flourishing. Delaney's chapter offers a Thomistic natural law perspective, presenting a view that has the potential to resonate with both religious and secular bioethicists. Delaney challenges traditional objections to enhancement, exploring how radical enhancements might expand access to existing basic goods and even introduce new ones. His nuanced argument bridges theological and philosophical approaches, advocating cautious engagement rather than outright rejection of enhancement technologies.

1.7.1 Delaney: Natural Law and Human Enhancement

Delaney grounds his analysis in the Thomistic natural law tradition, which conceives of human flourishing as the pursuit of incommensurable basic human goods, such as life, knowledge, friendship, and practical reasonableness. He applies this framework to assess the permissibility of enhancements, distinguishing between moderate enhancements (which expand human capacities within the existing spectrum) and radical enhancements (which create capacities beyond natural human limits).

Delaney addresses a central objection within natural law: that radical enhancements violate human nature. Critics argue that such interventions could fundamentally alter what it means to be human, replacing natural capacities with artificially engineered ones. Delaney counters this objection by distinguishing between changes that enhance human flourishing and those that disrupt it. He argues that enhancing capacities to access new goods need not constitute a substantial change to human nature, as long as the enhancements respect the principles of practical reason and avoid undermining existing goods.

A unique aspect of Delaney's argument is his exploration of the potential for enhancements to create access to entirely new basic human goods. He posits that just as art and justice emerged as distinct goods over the course of human history, radical enhancements could enable new forms of flourishing. For instance, enhancements might allow humans to communicate telepathically or achieve intellectual feats currently unimaginable. Delaney acknowledges the speculative nature of this claim but insists that the possibility warrants careful consideration.

This view challenges traditional Aristotelian notions of human flourishing as bounded by existing human capacities. Instead, Delaney adopts a more open-ended perspective, drawing on a Platonic view of goods as transcending human nature. By framing enhancements as a means of expanding human flourishing, Delaney offers a constructive response to critics who see enhancement as inherently unnatural.

Delaney approaches radical enhancements with a critical lens, emphasizing the importance of adhering to several guiding constraints to ensure their ethical implementation. First, he stresses the necessity of respecting basic goods, asserting that enhancements must not intentionally destroy or devalue existing valuable aspects of human life. For instance, while increasing cognitive capacities might seem desirable, such advancements should not come at the cost of undermining interpersonal relationships or moral virtues. Furthermore, Delaney highlights the importance of efficiency, arguing that enhancements should represent an effective and balanced means of pursuing human flourishing. Technologies that demand excessive resources or exacerbate social inequalities risk violating this principle by creating disproportionate burdens.

Central to Delaney's critique is the preservation of ontological integrity. He draws a distinction between enhancements that expand human capacities and those that fundamentally alter human nature. Enhancements that align with human rationality and embodiment, he argues, can contribute to human flourishing without transgressing the metaphysical boundaries of what it means to be human. In addition to these ethical considerations, Delaney incorporates a Christian perspective, addressing concerns about enhancement's potential to conflict with humanity's ultimate spiritual purpose: union with God. He warns against allowing an overemphasis on earthly goods to overshadow spiritual fulfilment, advocating for the good of religion to remain an "organizational priority" within the broader hierarchy of human goods.

Delaney's argument offers a rare synthesis of secular and religious bioethical perspectives. His reliance on natural law principles provides a framework that is accessible to non-religious bioethicists while remaining deeply rooted in Christian theology. This dual appeal makes his approach particularly valuable in pluralistic contexts, where debates about enhancement often polarize along ideological lines.

For secular bioethicists, Delaney's focus on expanding human flourishing aligns with humanistic ideals of progress and innovation. For religious bioethicists, his emphasis on respecting human nature and preserving the good of religion addresses concerns about the spiritual and moral implications of enhancement.

1.7.2 *Editor's Commentary*

Delaney's chapter offers a thought-provoking exploration of radical human enhancements through the lens of Thomistic natural law. By blending philosophical depth with theological insight, he crafts a framework that is both innovative and respectful of human dignity and flourishing. Delaney's argument that enhancements can expand access to existing basic goods, while also introducing new ones, challenges conventional boundaries of human potential. His proposal is grounded in an approach that prioritizes ethical constraints, such as the prohibition against the intentional destruction of existing goods and the principle of incommensurability among basic human goods. This foundation facilitates a vision for enhancement that is ethically robust and philosophically intriguing, though not without points of contention.

One of the chapter's notable strengths lies in its effort to reconcile secular and religious bioethics, demonstrating that enhancement technologies need not inevitably undermine spiritual fulfilment or human nature. Delaney's emphasis on humility and caution provides a counterpoint to the arguably uncritical optimism often associated with transhumanism. His work encourages a sophisticated engagement with enhancement, navigating between unwarranted fear and blind acceptance. However, his optimism about the compatibility of radical enhancement with natural law will not be accepted by all religious bioethicists.

The chapter raises a number of thought-provoking questions. How do we ensure that enhancements respect the unique value of all basic goods without diminishing any? How do we safeguard against the social and metaphysical risks of enhancement programs? Delaney's vision for new basic human goods also invites a critical examination of the boundaries of human nature and flourishing, as well as the plausibility of his claims regarding their ethical permissibility.

Addressing the overarching question—should religion have a seat at the table of bioethics—Delaney presents a strong case for its inclusion. The integration of Thomistic principles illustrates how religious perspectives can enrich bioethical debates, particularly by providing ethical and spiritual guidance in discussions on enhancement technologies. While his framing of enhancement as a potential means to deepen human flourishing rather than distract from ultimate spiritual ends is thought-provoking, it remains open to challenge. Delaney's chapter underscores the valuable role religion could play in bioethical discourse, though the extent to which his vision can withstand critical scrutiny remains a matter for debate.

1.8 Conclusions

The authors of the chapters in this volume examine the central question of whether religion should be permitted a seat at the table of bioethics, in the context of a range of specific topical issues in contemporary bioethics. It is evident that the positions of the secular and religious thinkers differ markedly in their conclusions as to the extent to which a seat should be pulled out for religion.

The secular authors approach the question of religion's place in bioethics with a commitment to public reason, arguing that ethical frameworks must be grounded in principles accessible to all, regardless of faith or lack thereof. They acknowledge that religious perspectives may offer insights into major ethical issues, but contend that these contributions must be reframed in universally comprehensible terms to justify their inclusion in bioethical debates. For these authors, religion's role is conditional: it can provide a voice but not a privileged seat. Any argument or perspective drawn from religious doctrine must withstand the scrutiny of rationality, universality, and coherence to earn a place in the public sphere. This insistence on secular norms reflects a broader concern that privileging religion risks undermining the inclusivity and adaptiveness needed to navigate the ethical challenges of today's pluralistic, liberal and increasingly secular societies.

By contrast, the religious contributors to this book assert that religion deserves a substantive role in bioethics, emphasizing its capacity to address questions of meaning, purpose, and ultimate moral grounding that, they claim, secular frameworks often leave unresolved. They argue that excluding religion risks marginalizing not only significant cultural and historical resources but also the lived experiences and values of religious communities. For these authors, religion's inclusion is not merely desirable but necessary to ensure a truly comprehensive bioethical discourse. They contend that religious traditions bring unique moral insights and long-standing reflections on human dignity and justice that can enrich public deliberation. However, they also at least acknowledge the importance of dialogue and the need for religious bioethicists to engage constructively with secular reasoning, seeking common ground where possible.

This divergence is particularly evident in how the two groups view the terms of engagement. Secular authors stress the primacy of shared rational principles as the baseline for any participant at the bioethics table, warning against sectarianism or the imposition of faith-based moralities. Religious authors, on the other hand, highlight the value of moral diversity and the purported limitations of a purely secular approach. They argue that religion offers a broader ethical lens, encompassing relational and existential dimensions often absent in secular bioethics, and that its voice should be respected as equal, even if not dominant.

Fundamentally, the question of whether religion should have a seat at the table of bioethics hinges on deeper disagreements about the foundations and purposes of bioethical inquiry itself. For the secular authors, the table is defined by public reason, and religion is welcome only insofar as it adheres to this standard. For the religious authors, the table must accommodate a plurality of moral voices, including those rooted in faith, recognizing that secular and religious perspectives can complement one another in grappling with the profound ethical dilemmas of our time.

Ultimately, the question of religion's role in bioethics hinges on two pivotal considerations: how we define 'religion' and the boundaries of bioethical discourse. Bioethics engages with profound moral questions surrounding medicine and healthcare, where religion's values often intersect with secular concerns. Critics argue that religion's reliance on unfalsifiable assumptions undermines rational debate, yet secular moral frameworks also rest on foundational assumptions that elude strict rational justification. Proponents of religion's inclusion emphasize its ability to represent the values of significant populations, enriching moral deliberations with diverse perspectives. Notably, none of the secular-oriented authors in this volume deliver an unequivocal 'no' to religion's inclusion; instead, they explore the question of *what form* such inclusion should take. In contrast, the religiously oriented authors advocate positions ranging from equal representation of religious and secular voices to privileging specific religions at the head of the table. Whether as equal participants or privileged guides, their debate reveals that religion's role in bioethics is less a question of presence and more a matter of placement—and the type of seat it occupies profoundly shapes the conversation.

Chapter 2
Rational Bioethics Without Religion

Jonathan Baron

Abstract Current bioethics reflects current local social norms, which are derived in part by reflective analysis of consequences but also from religious documents and traditional values seen as religious. Some of these norms would not withstand scrutiny based on actively open-minded thinking, which requires looking for reasons why a favored possibility might be wrong. We can, however, develop a system of reasoning to arrive at moral judgments in general based on three principles: benefit/harm, fairness to all affected, and rejection of claims that some principles are objectively true, whatever the consequences. Application of these principles leads to a form of utilitarianism. Thinking this way involves application of various helpful tools, such as the veil of ignorance, the generalization test, and thinking of interpersonal conflicts as if they were intra-personal conflicts between conflicting goals. These tools require role-taking ability, which also must be learned.

2.1 Introduction

Religious approaches to bioethics, or more generally to morality, involve the application of principles to cases, much like the law. In common law, the principles come from legal precedents written by judges, and from judges' interpretations of laws and constitutions. In religions, the principles usually come from interpretations of ancient texts (roughly analogous to interpretations by judges) or from unwritten social norms passed from one generation to the next (with occasional modifications).

For example, the Belmont Report (National Commission for the Protection of Human Subjects of Biomedical and Behavioral Research 1979) says, "The Hippocratic maxim 'do no harm' has long been a fundamental principle of medical ethics. Claude Bernard extended it to the realm of research, saying that one should

J. Baron (✉)
Professor Emeritus, Department of Psychology, University of Pennsylvania, Philadelphia, PA, USA

K. R. Smith (ed.), *Bioethics and Religion*, The International Library of Bioethics 116, https://doi.org/10.1007/978-3-032-33538-8_2

not injure one person regardless of the benefits that might come to others." This principle has many antecedents, but a major one is the Doctrine of the Double Effect (McIntyre 2023) which, as stated by Aquinas, attempts to define "injure" or "harm" more carefully so that unintended but necessary side effects are not considered harm. This doctrine has become a traditional part of Catholic moral theology (see Bennett 1966). When the decision involves a thumbs-up vs. thumbs-down choice, like whether or not to approve an abortion for a woman who would probably die without one (Hagerty 2010), the moral relevance of the distinction between "causing harm" and "failing to prevent harm" is unclear at best. These are the two options, whatever they are called. Yet the tradition in question counts harms induced directly through initiation of a physical chain of events as wrong but not harms induced by failing to block other causal chains. This widely employed distinction is justified by religious doctrine and is also part of the social norms of many cultures.

Bioethics as a field tries to separate itself from religion, but the influence of religion is inevitable because popular religions influence social norms of people outside of those religions. Consider the amount of attention in the bioethics literature given to human reproduction (abortion, in-vitro fertilization, embryo research, etc.) compared to animal rights. Some of the philosophical questions raised by abortion and animal rights are, arguably, the same. Yet animal-rights activists (including ethical vegetarians) usually do not see their concerns as arising from any religious doctrine.

Social norms are maintained and modified through the processes of cultural evolution (Cavalli-Sforza et al. 1982). As defined by Bicchieri (2006), social norms are beliefs about what behaviors are expected or discouraged. People are willing to try to enforce these norms on others in various ways (including gossip). Importantly, each person's adherence to a norm depends on that person's belief that others also adhere to it. Thus, norms can change over time depending on the frequency of adherence in an interacting population. And people also have beliefs about what norms should exist, independently of which norms exist in fact.

Much of bioethics is also an attempt to systematize social norms in written documents (such as the Belmont Report), but the norms continue to evolve. What is unacceptable to one generation, or one national culture, might be acceptable to the next generation or to a different national culture. For example, Hannikainen et al. (2023) present data showing that approval of euthanasia and assisted suicide increased steadily between 1981 and 2021 in 48 European countries, albeit at a faster rate at the eight countries that liberalized their law during this period (faster both before and after that change). These included Portugal and Spain, which are nominally "Catholic countries", where the dominant religion strictly bans euthanasia of any sort.

2.2 Thinking Without Religion

We can think about moral/bioethical issues without reference to social norms or authoritative documents. This is not to say that everyone should try to figure things out by themselves. Some people are overconfident in their ability to do this and often decide from "my own research" that purveyors of conspiracy theories and various eccentric religious sects have it right (Light et al. 2022).

Thinking on our own can sometimes work, if we think well. What does that mean? I have proposed a set of standards for good thinking, which I call "actively open-minded thinking" (AOT; Baron 2023). AOT is consistent with other efforts to define standards for thinking, although it is couched in the language of cognitive psychology (Baron 1985). It is based on a theoretical framework for all goal-directed thinking, from mathematical problem solving to foreign-policy decisions, to prediction of future events, to medical decisions by patients, providers or policy makers. Episodes of thinking may occur in seconds or years, and thinking can be done by individuals, groups, computers, or combinations of these. Thinking deals with *possibilities* (options), possible answers to the question that inspired the thinking. *Evidence* (arguments, reasons) strengthens or weakens possibilities under consideration. *Goals* (criteria, objectives, values) determine how the evidence is used. Thinking involves *search* for these three elements, and *inference* from the elements already found.

Consider a decision about a pregnant woman whose health, and possibly life, is threatened by continued pregnancy. Initially, the only goals might be to preserve the life and health of the mother and to deliver a healthy baby, and the initial options might be abortion and waiting. The evidence might come from prior studies or judgments of physicians about the degree of risk to the mother and the effect of waiting for a week, or two, on that risk. Other evidence might concern the feasibility of starting over with a new pregnancy if the fetus is aborted. A search for additional goals might yield the goal of conforming to the law, including precedents for malpractice suits. A search for additional possibilities might yield a cesarean section, and that would yield a search for evidence about the probability of the baby surviving in good health. The goals vary in importance, and the effect of each goal on the weight of each piece of evidence depends on the goal and the evidence.

The thinker keeps track of confidence that the currently favored possibility is the best one, and confidence in the efficacy of further thinking. Thinking should stop when confidence is sufficiently high and/or when the thinker believes that further thinking is futile. Thus, confidence in one's current thinking should be appropriate to the thinking done so far. And thinking must be fair so that it does not favor possibilities just because they are currently strong. This principle of fairness applies to both search and inference. If high confidence in the favored possibility results from failure to look for counter-evidence or alternatives, this confidence violates the standards of AOT.

To varying degrees, people favor the standards of AOT or oppose these standards. People tend to follow their own standards (Baron 1991, 1995; Stanovich and West 1997), and they use these standards to evaluate others (Baron 1993, 2019, 2024).

In the modern world, individuals do not know enough to think effectively about many decisions. We must rely on others to do our thinking for us: physicians, researchers, lawyers, financial advisors, newspaper columnists, and even politicians. This means that we must evaluate these potential sources, and the theory of AOT can be applied there too. Indeed, for many kinds of decisions, the evaluation of sources may be the main function of adherence to the standards of AOT.

In sum, the main idea is that we must look actively for reasons why tentative conclusions are wrong. When we find such reasons, we should treat them fairly. We may need to modify our conclusions. In other words, we should avoid "confirmation bias" (also called "myside bias") which is one of the major violations of AOT standards.

The other major violation of AOT is "uncertainty aversion" (Baron 2024). When we feel certain, we do not feel the need for further thought (or consultation with others). Yet certainty often results from self-deception: people convince themselves that they are certain when the quality and quantity of their thinking does not justify such certainty. And, when people confuse expressions of certainty with good thinking, they put more trust in sources who convey a sense of certainty, however unwarranted it may be. The feeling of certainty increases with more thinking even when the thinking is ineffective (e.g., Shynkaruk and Thompson 2006). Thus, people who are confident in their beliefs because they have spent a lot of time thinking about them may be deceiving themselves, especially if the thinking was affected by confirmation bias.

2.3 Thinking About Religion

The avoidance of confirmation bias alone can justify questioning of conclusions that come from religious doctrine. Some people who derive morality from religious principles may suffer from confirmation bias. They fail to present themselves with straightforward challenges.

One major challenge starts with the fact that religions differ, so it is reasonable to ask, "Why do you think your religion is correct and others are incorrect?" Of course, some people do not make this claim.[1] But it is difficult to avoid such a claim when people want the law and other institutions to follow their religious principles. Of course, many religious people do not think that their religious beliefs should

[1] During the psychedelic hippie period of the 1960s, a fellow graduate student was extolling the virtues of LSD and a form of Buddhism that advocated becoming "one with the all" as a means to world peace and brotherhood. Another student said, "You sound like a soapbox orator who says that the solution is for everyone to accept Jesus Christ as savior." The first student thought for a moment and then answered, "Yes, that would work too."

affect policies that they impose on others. (Famously, John F. Kennedy had to convince U.S. voters in 1960 that he would not let his Catholic religion dictate his policies as president.) Many citizens who oppose abortion on religious grounds do not think it should be illegal.

A counter-argument, which some religious people may make to themselves or to others who challenge them, is to ask whether any alternative to their religion is any better: "Atheism is just another religion. What makes you sure that your views are better than mine?" Although atheism is not so much another religion but a rejection of certain beliefs, we (both theists and atheists) can construct morality without relying on religion.

2.4 Where Does Morality Come From?

The answer does not depend on deriving a moral system from scratch. Morality is a cultural invention, developed over tens, perhaps hundreds, of thousands of years. We have written records going back a few thousand years. The invention of writing was surely an occasion for more rapid development, but some current cultures still lack such a tradition, particularly "uncontacted peoples", that is, groups of indigenous peoples that have tried to isolate themselves from contacts with the outside world, such as the Hadzda of Tanzania, and that lack formal schooling. These groups provide clues about earlier systems of morality (Hallpike 2004). Such early systems make no sharp distinctions among etiquette, religion, law, and morality. It was all "What we do." And the justification for the resulting social norms came from the fact that various people endorsed them, sometimes chiefs, and sometimes the group through general acceptance. Such concepts of morality continue to exist alongside of later-developing concepts in modern societies.

The modern idea of morality arose out of a search for ways to criticize accepted social norms, to change them, or to resolve conflicts between competing norms or systems. True morality is "post-conventional" in the sense of Kohlberg (1963), who classified norms based on law or social consensus, without additional justification, as "conventional." Any attempt to develop critical (post-conventional) moral principles must involve a willingness to put conventional norms aside. But it must also inherit two properties of conventional norms. First, it must take non-moral preferences into account. The idea of harming or helping other people is dependent on what they find harmful or helpful. Harm and help include, respectively, experienced pain and pleasure, but morality need not be limited to experiences. People value states like the existence of pristine wilderness, or the well-being of their heirs after they die, and, on some accounts, people are harmed if these values are subverted. If someone is siphoning money out of one of your accounts (which is too complicated for you to bother trying to keep track of), without your knowledge, surely you are harmed even if you never find out about it. The ideas of avoiding harm or providing help are included in essentially all social norms that are considered moral. In this respect, critical moral thinking is built on what already existed.

The second property, also common in conventional norms, is that critical moral thinking must be "fair" to the preferences of different people. When a person with red hair proposes a norm implying that people with red hair should be first in line, any child on the playground will say that this "isn't fair." People should be treated equally unless we have a good reason to treat them differently.

These two properties are minimal, yet part of the enterprise. Both concepts, benefit/harm and fairness, are more complex than they seem at first. Later I shall discuss a third property of moral judgments, which is that they are prescriptive, hence not evaluable as claims about truth or falsity. As is the case with other conceptual tools, such as logic, probability, and mechanics (as a part of physics), scholars have developed and elaborated a few basic properties, and their ideas have penetrated general culture to various degrees.

The first two properties, benefit/harm and fairness, can do quite a bit of work. Pure racism, which allowed people to justify enslavement of Africans, was vulnerable to critical arguments. What property of Africans justifies the neglect of their preferences? Candidate answers were easily dismissed. Likewise, discrimination against women eventually fell into line with empirical facts: the main difference is that men are statistically bigger and stronger (so that separation in many sports is justified). Women do somewhat better on the average in academic tasks, so discrimination in areas that require intellect is certainly not justified. The widespread historical dominance of men may have resulted from their taking advantage of their size and strength to suppress women physically, but this sort of inequality cannot withstand challenges as a moral principle, even though some religions still support inferior roles for women.

More generally, application of the impartial perspective required by fairness can lead to rejection of unfair norms. Various mental tools of argument have developed for such tests, starting perhaps with forms of the Golden Rule, which has been recognized for thousands of years by practically every religious and moral tradition. Some forms of that rule were limited to harms: do not do to others what you would not want done to you. Other forms were more general but somewhat imprecise and limited to undefined categories, e.g., "Love your neighbor as yourself" (Leviticus 19:18). Still others were limited to positive acts and failed to acknowledge differences in preferences, e.g., "Do to others what you would want them to do to you" (Matthew 7:12). Later interpretations of these early formulations became more consistent with the general idea that fairness in dealing with a single other person required taking into account the preferences of both people, and that this requires complete role taking. That is, proper consideration of other people requires seeing the situation as they do, with their preferences and limitations, not yours.

2.5 Modern Extensions of Benefit/Harm and Fairness

Other devices involve more than two people. The "veil of ignorance" (VOI) (Harsanyi 1955, 1975) asks us what decision we would make if we were behind a veil in which we did not know our own identity. We have an equal chance of becoming any one of those affected by the decision, like actors reading a play without knowing their roles. Of course, when we become another person we take on their preferences and limitations. In general, Harsanyi and others have argued that application of VOI reasoning to big questions about moral systems leads to a form of utilitarianism, in which more moral options are those that lead to a higher total (or average) utility (good, or goodness) over all those affected. This conclusion is based on the equivalence of "equal chance of becoming anyone" with "equal consideration of everyone."[2]

Huang et al. (2021) asked subjects to apply VOI reasoning to a problem in bioethics, the allocation of potentially life-saving ventilators in the early stages of the COVID-19 pandemic. The question was whether younger people, with a longer life ahead of them, should get priority over older people. Subjects were more likely to favor younger people when they were asked to apply VOI reasoning than when they were not. The effect of VOI reasoning was especially large among older subjects, who tended to resist the idea that younger people would get priority in the absence of VOI reasoning.

Although the VOI argument can be effective in experiments, it does not seem to be so familiar to philosophically unsophisticated people, perhaps because it requires a greater stretch of imagination of a counterfactual situation. Just what does it mean to take on someone else's identity? And the idea of making decisions in terms of expectation based on probability is not always so intuitive (Baron 2023, ch. 11).[3]

Hare (1963, 1981) suggests another device (which I shall the "ideal observer"). We make personal decisions all the time that involve conflicts of preferences within ourselves. We consider the strengths of preferences and the extent to which they are satisfied by various options. The idea is that we take on the preferences of others, as if they were our own, and resolve conflicts the same way we ordinarily do. This can get complicated when many people are involved, but for many situations it is obvious what the result would be, as in the case of the Huang et al. study about ventilators just described. (The preferences of the younger patients would count more

[2] When people are subdivided into groups, a further requirement is that the average utility (balance of costs and benefits) of people in each group must be multiplied by the size or the group; the size of the group corresponds to the probability of being in that group.

[3] Note that, for Rawls (1971), the VOI did not lead to the utilitarian solution. He thought that the probability of being in the least advantaged group was "unknown," so that "Social and economic inequalities are to be arranged so that they are... to the greatest benefit of the least advantaged" (Rawls 1971, p. 83)." This argument has several problems (Harsanyi 1975; Baron 2023, ch. 11; Huang et al. 2019). For example, "greatest benefit to least advantaged" is dependent on the definition of the larger group. If we apply the principle to one nation, and then add others so that the group becomes larger, the least advantaged group may change (Ubel et al. 2001).

because they have longer to live.) Hare also argues that his device, generally applied, leads to utilitarian solutions. As Hare put is (Hare 1981, section 6:2):

> … if I have full knowledge of the other person's preferences, I shall myself have acquired preferences equal to his regarding what should be done to me were I in his situation; and these are the preferences which are now conflicting with my original prescription. So we have in effect not an interpersonal conflict of preferences, but an intrapersonal one; both the conflicting preferences are mine. I shall therefore deal with the conflict in exactly the same way as with that between two original preferences of my own.
>
> Multilateral cases now present less difficulty than at first appeared. For in them too the interpersonal conflicts, however complex and however many people are involved, will reduce themselves, given full knowledge of the preferences of others, to intrapersonal ones.

Note that Hare did not intend this device as something to be used from day to day to resolve moral questions. Rather, as I shall discuss later, Hare makes a distinction between intuitive moral reasoning and critical moral reasoning. Intuitive reasoning is what we rely on most of the time. Critical reasoning could be attempted in difficult cases or in justifications of policy, but the concept of critical thinking functions more as a demonstration that moral dilemmas could, in principle, be resolved through rational thought.

I also note that Hare's approach here is disputed by those who think that utilitarianism ignores the "separateness of persons". Brink (1993) explains what this claim might mean and why the idea is nearly incoherent. To paraphrase and simplify Brink's argument, he takes the claim to mean that, while it is reasonable for people to trade a lower quality of life in one part of their life (e.g., saving instead of spending) for a substantially higher part later (spending the extra money earned from interest), it is not reasonable to make this kind of trade-off between people, asking one to have a lower quality of life so that someone else can have a substantially higher quality, e.g., taxing the rich (who thus have a reduction) to help the poor (a much greater increase in well-being for a similar amount of money). If we (as policy makers) cause a loss in one person in order to provide gains to others, we must fully compensate the loser. If the tax is put into effect, then we must compensate the rich.

But note that, if the tax is repealed, we must compensate the poor. Thus, when the status-quo exists, it affects who is the winner and who is the loser. It is unclear why the status-quo should matter. Likewise, if the tax is considered but *not* enacted, then the poor are the losers from this omission, even though they are not losers relative to the status-quo. Thus the act/omission distinction matters as well as the status-quo. The only way to avoid these problems is to require unanimous consent (a social contract) for any policy change. If we assume that people will not consent to uncompensated losses, the policy change would have to help some and hurt nobody. Such changes, which would of course also be consistent with utilitarianism, are hard to find, and it is not clear why we should limit ourselves to just these.

The main idea behind Hare's device is to resolve conflicts of conflicting preferences of all those affected, no matter whose preferences they are, by aggregating them as if they were all my own. In making a judgment from my perspective as one of the affected parties, I must consider (my best unbiased judgment of) the preferences of everyone else. And I should do this if I were someone else and come up with the same answer about what should be done. (If I have been unbiased, then I must assume that everyone else has the same perceptions of everyone else's preferences. I can't do better than that.)

This sort of aggregation across people does indeed ignore the "separateness of persons," but it seems impossible to develop a system that prevents such aggregation. Hare's approach may seem disturbing and non-intuitive to some, but it seems to be the only way to develop a formal system that maintains the criterion of universalizability (in combination with prescriptiveness), which is essential for our concept of "moral." This is a theoretical device, a standard of what counts as a moral judgment. It is not intended for everyday use by real people. (More about this later.) On the other hand, an understanding of the basic idea may help real people engage in role taking, the most difficult part of moral judgment for real people.

Why aggregate this way? If any other moral principle or method yields a different conclusion about what should be done, then it will violate some preference without compensating fully by satisfying some other preference. In personal decisions with conflicting preferences, nobody would rationally choose an option that reduced the satisfaction of one preferences without making up for it somewhere else. These comparisons across preferences have already been made in determining the best option. So why would we adopt a principle that makes things worse?

Note that, in this scheme, opinions about moral principles are counted as preferences, but that is all. They must be weighed against other preferences. For example, if the thought that abortions are being done has a negative effect on you, such as a feeling of moral repugnance, then we must count your preference not to have this negative feeling as one of your preferences, which we trade off with the preferences of others. What we must not do is trade off your repugnance with the conclusion of an analysis based on preferences, even if you think we should do that. This is because your repugnance has two parts. One part is a preference that abortions not be done, which is much like other preferences—such as "Men should wear neckties in the office"—except perhaps in strength. That is what we count as a preference when we make trade-offs as an ideal observer. The other part is your belief that abortions should not be done even when an aggregation of relevant preferences, including yours, implies that an abortion would be best. You want to bring in a moral principle outside of the analysis based on preferences. We might ask where this principle comes from and why we should follow it.

Note that the preference part (as opposed to the external principle) is morally relevant, just as other preferences are. If enough people prefer that abortions should not be done, they could still determine the optimal outcome of a decision about laws covering a region where this was true. Likewise, in some Muslim countries, the vast majority adheres to principles of female modesty. Immodesty can thus harm these people by going against their preference even in the absence of any external principle. Immodesty that is acceptable in some other place might be morally unacceptable here.[4]

Returning to the case of the older and younger patients, suppose someone says, "I don't care about what would be chosen by an ideal observer. The principle of equality is overriding, so the older and younger patients must be treated equally." Following such a principle would indeed make things worse for the young without sufficient compensation for the old. A critical response (as prescribed by AOT) would lead us to ask why we should accept and follow such a principle as anything but a personal preference. The principle of prescriptiveness implies that it cannot be a claim about matters of fact, as I discuss shortly.

On the other side, consider what it would mean to endorse a moral principle that would give priority to younger patients. Such a principle would apply to the current situation, in which we know nothing about the patients except their ages, and the statistical fact that the life expectancy of the younger patients is greater. If we knew other things, they might be morally relevant. The principle we endorse applies only to cases like the present case in all morally relevant respects, which include age and the absence of other potentially relevant information.

The principle of fairness applied to moral judgments implies that the identity of the one endorsing the principle is *not* a morally relevant feature of the situation. To make an honest endorsement, as an old person, I must fully represent to myself not just my own preferences but those of an otherwise equivalent young person. The conflict I must resolve is thus, approximately, between living another 10 years (say) and living another 40 years. If I honestly represent the situation to myself, it is difficult to avoid the conclusion that the younger person should get priority.

2.6 Prescriptiveness

Hare (1952, 1963, 1981) has argued that moral judgments are not propositions about facts. They are imperatives, that is, prescriptions about what to do. A syllogism like "X is false; therefore don't say X" is incomplete. The fact of X being false does not imply that you shouldn't say it. "Don't say X" is an imperative. And an imperative premise is required to reach that conclusion, such as "Don't say what is false." Hare's argument amounts to logical analysis of "'Is' does not imply 'ought'." Similarly, "The Bible says not to do X; therefore do not do X." leaves us feeling (correctly) that something is missing. Why should I do what the Bible says?

Suppose someone says, or thinks, "I understand how you reach this conclusion, but the principle of equality says that we must treat people equally, so we must give

[4] Importantly, when we aggregate people's preferences, we serve them better if we ignore preferences based on false beliefs. For example, in deciding whether to give a life-saving injection to a patient, we might ignore the patient's resistance based on a belief that injections contain microchips that provide information to Bill Gates.

old and young an equal chance."[5] The statement that "the principle of equality says…" seems to be a factual assertion about what it says, not an imperative. Thus, the "so we must give old and young an equal chance" does not follow, since it is an imperative. The factual assertion is thus a pseudo-fact.

Controversies often turn on questions about which pseudo-facts are true. In the abortion debate, for example, those on one side sometimes claim that killing a fetus is absolutely forbidden, and those on the other side sometimes claim that women have an absolute right to control their "own bodies" (which begs the question of who owns the fetus's body). These are (sometimes) understood as absolute principles outside of any utilitarian calculations, thus applicable "whatever the consequences" (Bennett 1966). What can people who endorse one absolute principle say to those who endorse a conflicting absolute principle? Unlike many other matters of truth, this is not a matter that empirical investigation can ultimately resolve. Is there a book with the answer in it? or an oracle we can consult?

Assertions of such pseudo-facts are unlikely to survive actively open-minded thinking, since the evidence for them does not seem to exist.

2.7 How People Actually Think

The ventilator study of Huang et al. may illustrate another interesting feature of our moral judgments. The argument favoring younger patients is hard to refute, yet many people, when they realize this, will feel a certain repulsion, a mental equivalent of "ouch." This is because the argument conflicts with an intuition that most of us have favoring equal treatment in this case (and many others). The principle of equal treatment may be part of a social norm, held more strongly in some cultures than others (e.g., Murphy-Berman et al. 1984).

Some social norms can be justified as based on principles consistent with utilitarian thinking in many situations. When no other factors are relevant, equal division of goods usually maximizes total utility. If we start from equal division and then take from some in order to give to others, the utility gain for the gainers is usually smaller than the utility loss for the losers. (In the extreme case, if everyone has exactly a million dollars in the bank, the gain of a second million would be nice, but the loss of a million would be very bad.) Some people may be aware of the justification, or some version of it, and such awareness may help to maintain the norm. If norms are consistent with utilitarian reasoning most of the time, use of the norm to make decisions will lead to choices that are consistent with utilitarianism most of the time.

Other social norms that develop outside of any reflective system, e.g., religious systems that depend on authority or tradition rather than reasoning that looks for

[5] Such an assertion may seem implausible in this case, but it is found in other cases that involve strict equality vs. maximizing lives saved, e.g., Ubel et al. (1996, 2001).

possible alternatives and arguments. These include many beliefs about abortion (on both sides), euthanasia, homosexuality and gender change, or female conduct and roles.

Hare (1981) proposed that actual moral thinking has two levels, a critical level and an intuitive level. The critical level for Hare is what I have illustrated with the three principles and the ideal-observer approach, and I will assume that here.[6] The intuitive level consists of principles that we apply every day, like: it is bad to break promises; lying for self-interest is bad but white lies to protect others are OK; assume that people you deal with have good intentions unless you get evidence to the contrary. In medicine, other examples are: let people make their own decisions insofar as possible; in extreme cases is OK to intervene in order to protect people from themselves; avoid putting others at risk if possible; (for some) and don't try to save the insurance company's money by "bedside rationing" (Ubel 2000). Intuitions like these are the tools of everyday moral thinking, discussion, and debate.

Many intuitions work well in approximating the results of critical moral thinking. But intuitions may also arise from other sources, and some of these may lead to conclusions that are often inconsistent with a utilitarian analysis (hence lead to more harm than good).

Moreover, even when intuitions are justified by critical thinking, and also when they are not, they may take on a life of their own, so that they control judgments and decisions, or at least assert themselves as "ouch" responses, regardless of the critical analysis of consequences. Intuitive principles may also come to be understood as absolute, so that they apply regardless of the consequences (Baron and Spranca 1997). I call them protected values (PVs) because people see them as protected from trade-offs with other values. Most PVs are prohibitions of actions such as "Using genetic engineering to make people more intelligent." If too many of them were injunctions to act, they would often conflict with each other (Ritov and Baron 1999; Baron and Ritov 2009).

Baron and Leshner (2000) found that most PVs yielded to simple challenges of the sort an actively open-minded thinker would pose, e.g., "Can you think of a situation where this [prohibited] action would be acceptable?" Despite the apparent fragility of PVs in the face of AOT, people are known to act on these values, e.g., failing to rescue girls in a burning school because they were not properly veiled (Eltahawy 2002) or trying to prohibit an abortion that would save the mother when both mother and fetus would likely die without one (Hagerty 2010). Prior instruction in AOT might be able to prevent the development of these sorts of PVs.

Intuitions that are both strong and detached from their justifications can distort thinking. Given that some of these detached intuitions come from social convention and religion, thinking may be more successful in those people who have rejected, or put aside, those influences. This sort of rejection may result from an earlier step of actively open-minded thinking, in which people question their current beliefs. They

[6] This two-level distinction does not depend on the critical level being utilitarian. The same distinction could apply to any highly systematized moral theory.

reason that these beliefs are the result of how we were raised, and other people were raised differently. This is a "debunking" argument (Greene 2013), an argument showing that our beliefs may have some other cause than their correctness; such an argument does not imply that our beliefs are false. It could turn out that we were just lucky to be taught the correct beliefs while others were unlucky to be taught other beliefs. But the possibility that our beliefs have some other cause encourages us to look for alternatives.

So far I have tried to show how AOT can lead us to question principles that are grounded in authority or social consensus. And AOT is consistent with the sort of careful and extensive role-taking that is required for any approximation to the ideal-observer standard. AOT leads to ask how others see a proposed course of action when their values and concerns are different from ours. It is not a hopeless dream to think that AOT could increase in various populations. It is a matter of cognitive style, not capacity or skill (although its efficacy may sometimes depend on these), so it is under control. And it is a matter of degree, with many people showing fairly high levels of AOT acceptance.

It turns out that utilitarian reasoning seems to be consistently correlated with AOT acceptance. That is, people who are high in AOT acceptance are more likely to give utilitarian responses to moral dilemmas and other questionnaire items. For example, Patil et al. (2021, especially Figure S2-6 in the supplement)[7] found a correlation between a scale measuring endorsement of AOT standards and responses to sacrificial dilemmas, in which one person must be harmed or killed through an emotionally aversive act, such as breaking his bones, in order to save many, such as preventing the detonation of a bomb in a public place.

Baron et al. (2015), Study 4)[8] also found correlations between a version of the AOT scale and responses to two types of moral dilemmas and one measure of acceptance of utilitarian reasoning. One type of dilemma, like those used by Patil et al., put action on the utilitarian side and numbers of harms on the deontological side.

The other type put affective appeal, usually through sympathy, on the utilitarian side, action, but an important rule on the other side, e.g.:

> X is doing research on depression in college students…She has promised the human subjects committee,…, to use code numbers for her subjects, so that she does not learn anything about their identity. A computer program assigns the code numbers, and then puts all identifying information in a separate data-base. One of her subjects writes a comment on a questionnaire, saying that he plans to kill himself tomorrow.
>
> Despite her promise, X knows how to write a computer program to recover identity information from the code numbers. What should X do?

[7] The text of the paper presents this result, the only one that uses a "classic" utilitarianism measure (categorizing all responses as utilitarian or not, as done in most other studies of moral judgment), as part of a regression model, leaving open the possibility that it arose as a suppression effect.

[8] References to this study apply to the preprint version at https://psyarxiv.com/g5jhp, which presents results in more detail. Most references also apply to the published version, which is available electronically in some libraries.

Recover the student's identity and contact someone close to the student. The suicide will be prevented.

Keep her promise and do not search for the student's identity. The student will commit suicide.

The six-item utilitarianism scale used by Baron et al. (2015) was as follows:[9]

- When a moral rule leads to outcomes that are worse than those from breaking the rule, we should **follow** the rule.
- When a moral rule leads to outcomes that are worse than those from breaking the rule, we should **break** the rule.
- When two options harm other people in the same ways, we should choose the option that harms fewer people.
- When some action causes harm to some people but prevents the same harm to many more people, we should act.
- It is morally wrong to harm some people in order to prevent the same harm to more people.
- Sometimes we should follow moral rules that prevent us from doing what is best on the whole.

This scale had low reliability (alpha of.50); the items did not correlate highly with each other, but the lower correlations did not follow any obvious pattern. Correlations of AOT with three very different measures cannot be explained in terms of reliance on affect.

We also included a "Religion" scale, consisting of six items[10] from Piazza and Sousa (2013): Response scale: Agree...Disagree (four points):

- The truth about morality is revealed only by God.
- It is possible to live a righteous life without knowledge of God's laws. (−)
- Acts that are immoral are immoral because God forbids them.
- We don't need to try to figure out what is right and wrong, the answers have already been given to us by God.
- An atheist can still understand what is morally right and wrong. (−)
- Without God, humans still have a way to distinguish right from wrong. (−)

Belief in "divine-command theory" (DCT) assessed by this scale is consistent with a belief that it is not in our power to question God's word. We might expect that people who have this kind of belief have low acceptance of AOT, which they might associate with liberalism, secular humanism, and agnosticism. If so, they should get low scores on the AOT scale, as found by Piazza and Landy (2013).

AOT correlated with all three measures of utilitarian thinking, and negatively with the Religion measure (DCT, in this experiment). In essence, everything correlated with everything else, as shown in Table 2.1, as if they were all affected by some common factor. Baron et al. suggested the term "cognitive liberalism". (The

[9] This was incorrect in the published version.

[10] Incorrectly stated as four items in the paper.

Table 2.1 Correlations, disattenuated correlations, corrected for imperfect reliability, above diagonal, raw correlations below it, Study 4 from Baron et al. (2015). CRTrt is the mean log response time on the CRT items. Uscale is the full utilitarian-belief scale. Rule and Number are the dilemmas. Relig is the 4-item religion scale

	Relig	AOT	Uscale	ActRule	ActOmit
Relig	**0.83**	−0.817	−0.808	−0.264	−0.346
AOT	−0.609	**0.67**	0.683	0.417	0.285
Uscale	−0.570	0.433	**0.60**	0.318	0.611
Rule	−0.200	0.284	0.205	**0.69**	0.436
Number	−0.270	0.200	0.404	0.310	**0.73**

same measures correlated negatively with political conservatism, but that was not measured in this study.)

Where do these correlations (and other similar ones described by Baron et al. 2015) come from? One possibility is that they result partly or fully from the way subjects answer questions in psychology experiments. Specifically, people who take time to reflect on questions may be more likely to make utilitarian responses. Yet, beliefs in DCT, the existence of God or gods, and conservative politics are surely acquired before doing any psychology experiment. Moreover, several experiments cast doubt on the role of reflection in accounting for individual differences in utilitarian moral judgment. For example, Bago and De Neys (2019) forced subjects to give a quick, intuitive response, followed by a reflective response to the same dilemma. Intuitive utilitarian responses were essentially just as frequent as reflective ones. Other results show that switches occur equally in both directions while subjects are deliberating about a single dilemma (e.g., Koop 2013) and that deontological responses take as much time as utilitarian responses (Baron and Gürçay 2017). Individual differences in utilitarian responding thus do not seem to depend on different approaches to experiments. They seem to result from the life histories of the subjects.

Baron et al. (2023) attempted to find the source of the correlations between AOT and utilitarian responding in life histories. Did AOT cause people to become more utilitarian? Did common social influences affect both (e.g., family background, higher education)? Did people just come to utilitarian reasoning, and AOT, on their own? To provide a preliminary answer to questions like these, we asked adult subjects to remember their past, and the views of those who "brought them up." We asked several questions about religious and political beliefs, and about their acceptance of utilitarianism and of the standards of AOT. Here is an example, using one of the "religion" questions other than those concerned with DCT:

God, and gods, do not exist, despite what people believe.

Completely disagree 1 2 3 4 5 Completely agree

How would you have answered this question earlier in your life, when you first might have been able to understand the question as you do now?

Completely disagree 1 2 3 4 5 Completely agree

How would those who brought you up have wanted you to answer this question?

Completely disagree 1 2 3 4 5 Completely agree

Please indicate how each the following influenced on your views.
I thought about things like this myself.
[Strongly toward 'disagree'...Strongly toward'agree']
People involved in my upbringing.
Other people I met later.
My formal education (e.g., school, college).
Specific life experiences that I had.

As shown in the first three questions of the example, subjects answered each for their current beliefs, their past beliefs when they could first think about these issues (presumably in adolescence), and their upbringing. Comparison of present and past indicated that they became more "cognitively liberal"—favoring AOT, less religious, more utilitarian, but not more politically to the left—over their lives. These changes were not highly correlated with each other, suggesting that no one of them caused the others. Specifally, changes in AOT did not correlate with changes in other domains. But note that each of these domains may develop piece by piece, domain by domain. The subjects attributed their changes to influences from others (teachers, friends, etc.) and their own thinking.

Answers to the question about influence of upbringing ("People involved...") correlated with the past-present change in each domain. For example, upbringing-influence questions about AOT correlated withe the past-present change in AOT. However, these correlates were near 0 for effect of upbringing-influence across domains, e.g., upbringing-influence for AOT had no relation to past-present difference in Religion. The only exception to this lack of relationship was Religion and DCT, where the influence was strong in both directions.

The influence questions were separate: subjects could indicate "strongly" for more than one. All measures (averaging the scale scores) correlated positively with current answers, although the correlations of upbringing were weaker than the others. And all measures correlated positively with change from past to present, except for upbringing, which correlated *negatively* with change, as if those who changed the most were also the most rebellious. Otherwise, the strongest positive influences overall were for "I thought about things like this myself" and "specific life experiences..."

In sum, no particular category played a large role in affecting broad changes; in particular, AOT did not. When subjects indicated that changes resulted from their own thought, they seem to have meant that they thought about each issue.

Our measure of utilitarianism was particularly poor because lay beliefs in utilitarianism are usually partial, accepting some implications but not others. For example, long before I heard of "utilitarianism", I had already begun to understand that "two wrongs don't make a right," so that punishment could be justified by deterrence but not by retribution. Punishment was not even part of our measure, which focused on trade-offs across people (hurting some to help others) and on the conflict between rules and consequences.

The AOT scale itself may not be the best measure of acceptance of the standards of AOT *in practice*. Rather, it seems possible that AOT itself is something that people discover for themselves. The starting point may be the simple recognition that people make mistakes, and that one way to prevent such mistakes is to look for reasons why some conclusion of the moment could be wrong. People can of course learn to do this sort of search without recognizing what they are doing, but our scale requires the next step to a meta-cognitive principle, which then becomes an explicit standard. The scale measures acceptance of looking for negative reasons as meta-cognitive abstraction. Yet such abstractions are not so alien to everyday thinking: we have expressions in common language that arise from this sort of meta-cognition, e.g., "pig-headed", "narrow-minded", and "open-minded". The next step after this understanding of myside bias is the realization that overconfidence can shut down search prematurely: "It ain't what you don't know that gets you into trouble. It's what you know for sure that just ain't so."[11] An alternative approach to measurement of AOT acceptance in practice is to ask for judgments of other people's thinking, as done, for example, by Baron (1995, 2019).

In fact, some people become more cognitively liberal over their lives. Some of this is the effect of "thought about it myself." Such thought seems to be affected by other influences aside from the sort of upbringing we asked about, such as teachers, friends, or experiences with strangers. But it does not seem to be simply the transmission of culture, as that would not explain the apparent movement toward cognitive liberalism.

2.8 Conclusion

The concept of a sort of morality that is beyond social convention is fairly new in human history. This is the kind of "post-conventional" morality that we need to criticize and improve the standards of conduct consisting of "what we do" or "what the authorities tell us we must do." Like other recent concepts, such as probability, overall well-being (utility?), and time (as the object of arithmetic), old concepts persist alongside the new, and some versions of the new forms are conceptually incoherent. Hare's ideal-observer approach starts with basic criteria of what morality must be, namely, that it is prescriptive and cannot be derived from facts alone, and that it is universal in the sense that it is "unfair" to advance a moral principle that gives special status to the person doing the advancing. I have sketched why I think we should accept this approach as the best solution to the problem of defining morality in the sense desired.[12]

Conflicts between religion and any abstract post-conventional moral system are common, but actively open-minded thinking, plus ideal-observer theory, can help to

[11] Attributed to Mark Twain but not found in any of his writings.

[12] I confess that my intellectual style is to look for and defend solutions rather than looking for problems without solving them. I am bothered by common statement in the moral-judgment literature that "experts don't agree" so that, therefore, we can do purely descriptive research without any assumptions about the right answers to moral questions.

deal with these conflicts. Religions of the traditional sort rely on moral systems backed by appeal to authority, sometimes in the form of written documents.[13]

I have argued that a form of rational thinking, characterized by the standards of AOT, can lead to conclusions with bioethical significance. In Baron (2006) I have spelled out a number of other examples, such as those concerning competence to consent to participation in research, and have suggested some solutions. But that book did not include much discussion of the justification of utilitarian thinking, which I have briefly sketched here.

Schools at all levels could promote AOT and utilitarian thinking. Many schools now give general guidance about writing, sometimes in the form of specific principles to learn and remember, such as those in later editions of Strunk and White (e.g., Strunk Jr. and White 1979). Mathematics courses often point to classes of useful procedures such as indirect proof, mathematical induction, or testing a proposed general solution with simple examples. The same could be done with the principles of AOT and with the three procedures for critical thinking about morality. These devices can be taught simply as helpful devices for thinking, like mathematical induction. Thus, teaching them is not the same as indoctrination in utilitarianism.

The most difficult part of any such instruction will surely involve getting students to fully take the role of other people affected by a decision. It could be a great advance if more people learned simply to understand the perspective of someone else with different values, capabilities, and personal limitations.[14] Baron et al. (2018) review and criticize results claiming to show that such perspective taking (as it is often called) is negatively correlated with utilitarian reasoning, and we present some weak results showing positive correlations.

"Empathy" is often seen as involving two major components. One is literally "I feel your pain," a negative emotional response to learning of the suffering of others. The other component is a matter of understanding what it is like to experience suffering (or joy), but without experience the corresponding emotions yourself. It is understanding or role taking, the second component, that is required for moral reasoning. We must represent to ourselves the pleasure and pain of others so that we can weigh them appropriately.

A second major impediment to utilitarian thinking in the real world is parochialism, a limitation of moral concern to a particular in-group, such as citizens of a nation. Parochialism contrasts to cosmopolitanism, a concern for all humanity (or possibly all sentient beings). Political campaigns often appeal to the parochial voter (even more so when they involve "identity politics"), so people may get the

[13] From my own limited experience, clear example of religions *not* of the traditional sort are Unitarianism (U.S. version) and Quakerism. Other religious groups often have internal conflicts, such as Reform Judaism.

[14] I was once trying to get a successful businessman to understand redistribution. I asked him to imagine that he could either be himself or the homeless beggar on the street corner with equal probability (a coin flip) and then think about whether he should be taxed to help the beggar find a better place to sleep. He said that, if he were the beggar, he would pull himself up by his bootstraps and again become successful.

impression that cosmopolitan thinking is rare. It is not, although individual differences are large (Buchan et al. 2009, 2011; Enke et al. 2023).

Where do religions stand on these major issues? I have been unable to find any good studies of correlations between (e.g.) religiosity and role taking, or religiosity and cosmopolitanism, although the little that is available shows very low correlations either way, at best. That said, it is noteworthy that the strongest and most influential recent advocate for cosmopolitanism was the late Pope Francis. And some of the non-governmental organizations (NGOs) doing the most to help people with the most extreme suffering are religious. Roser et al. (2022) collected a series of articles about the relation between these NGOs and the effective-altruism movement, which tended to emphasize secular charities from the outset. Religious NGOs sometimes reject the utilitarian philosophy of the effective-altruists, and they often disagree on specific policy issues, such as abortion and family planning. Yet the areas of agreement are large, and the total financial support for religious NGOs is also much too large for those who want to relieve suffering in the world to ignore. Compromise should be possible.

I have argued that morality can be, and should be, rationally derived without help from religion. This conclusion does not imply that religion is "wrong" in other respects. Indeed, when I contemplate the birth of a child—a miracle witnessed almost daily by obstetricians and midwives—I am moved to a sense of mystery that must lie behind the impulse to believe in some religion (which may still be giving hasty answers to an unanswerable question). The mystery is well captured by these excerpts from *The salutation*, a poem by Thomas Traherne (ca. 1637–1674):[15]

> These little limbs,
> These eyes and hands which here I find,
> These rosy cheeks wherewith my life begins,
> Where have ye been?...
> The earth, the seas, the light, the day, the skies,
> The sun and stars are mine if those I prize.
> ...
> A stranger here
> Strange things doth meet, strange glories see;
> Strange treasures lodged in this fair world appear,
> Strange all and new to me;
> But that they mine should be, who nothing was,
> That strangest is of all, yet brought to pass.

[15] Set to music by Gerald Finzi, in *Dies natalis*.

References

Bago, B., and W. De Neys. 2019. The Intuitive Greater Good: Testing the Corrective Dual-Process Model of Moral Cognition. *Journal of Experimental Psychology: General* 148 (10): 1782–1801.

Baron, J. 1985. *Rationality and Intelligence*. New York: Cambridge University Press.

Baron, J. 1991. Beliefs About Thinking. In *Informal Reasoning and Education*, edited by J. F. Voss, D. N. Perkins, and J. W. Segal, 169–186. Hillsdale, NJ: Erlbaum.

Baron, J. 1993. Why Teach Thinking? — An Essay (Target Article With Commentary). *Applied Psychology: An International Review* 42:191–237.

Baron, J. 1995. Myside Bias in Thinking About Abortion. *Thinking and Reasoning* 1:221–235.

Baron, J. 2006. *Against Bioethics*. Cambridge, MA: MIT Press.

Baron, J. 2019. Actively Open-Minded Thinking in Politics. *Cognition* 188:8–18.

Baron, J. 2023. *Thinking and Deciding*. 5th ed. New York: Cambridge University Press.

Baron, J. 2024. Two Components of Individual Differences in Actively Open-Minded Thinking Standards: Myside Bias and Uncertainty Aversion. Working Paper. https://sjdm.org/~jbaron/ms/ws/ws.pdf.

Baron, J., and B. Gürçay. 2017. A Meta-Analysis of Response-Time Tests of the Sequential Two-Systems Model of Moral Judgment. *Memory and Cognition* 45 (4): 566–575.

Baron, J., and S. Leshner. 2000. How Serious Are Expressions of Protected Values. *Journal of Experimental Psychology: Applied* 6:183–194.

Baron, J., and I. Ritov. 2009. Protected Values and Omission Bias as Deontological Judgments. In *Moral Judgment and Decision Making, Vol. 50 in B. H. Ross (series editor), The Psychology of Learning and Motivation*, edited by D. M. Bartels, C. W. Bauman, L. J. Skitka, and D. L. Medin, 133–167. San Diego, CA: Academic Press.

Baron, J., and M. Spranca. 1997. Protected Values. *Organizational Behavior and Human Decision Processes* 70:1–16.

Baron, J., S. Scott, K. Fincher, and S. E. Metz. 2015. Why Does the Cognitive Reflection Test (Sometimes) Predict Utilitarian Moral Judgment (and Other Things)? *Journal of Applied Research in Memory and Cognition* 4 (3): 265–284.

Baron, J., B. Gürçay, and M. F. Luce. 2018. Correlations of Trait and State Emotions With Utilitarian Moral Judgments. *Cognition and Emotion* 32 (1): 116–129.

Baron, J., O. Isler, and O. Yilmaz. 2023. Actively Open-Minded Thinking and the Political Effects of its Absence. In *Divided: Open-Mindedness and Dogmatism in a Polarized World*, edited by V. Ottati and C. Stern. Oxford University Press.

Bennett, J. 1966. Whatever the Consequences. *Analysis* 26:83–102.

Bicchieri, C. 2006. *The Grammar of Society: The Nature and Dynamics of Social Norms*. New York: Cambridge University Press.

Brink, D. O. 1993. The Separateness of Persons, Distributive Norms, and Moral Theory. In *Value, Welfare, and Morality*, edited by R. G. Frey and C. W. Morris, 252–289. New York: Cambridge University Press.

Buchan, N. R., G. Grimalda, R. Wilson, M. Brewer, E. Fatas, and M. Foddy. 2009. Globalization and Human Cooperation. *Proceedings of the National Academy of Sciences* 106:4138–4142.

Buchan, N. R., M. Brewer, G. Grimalda, R. Wilson, E. Fatas, and M. Foddy. 2011. Global Social Identity and Global Cooperation. *Psychological Science* 22:821–828.

Cavalli-Sforza, L. L., M. W. Feldman, K. H. Chen, and S. M. Dornbusch. 1982. Theory and Observation in Cultural Transmission. *Science* 218 (4567): 19–27.

Eltahawy, M. 2002. They Died for Lack of a Head Scarf. Washington Post, March 19. https://www.washingtonpost.com/archive/opinions/2002/03/19/they-died-for-lack-of-a-head-scarf/86cbcc5a-6cef-49e5-b758-167527e6eed0/.

Enke, B., R. Fisman, L. M. Freitas, and S. Sun. 2023. *Universalism and Political Representation: Evidence From the Field*. Vol. No. w31265. National Bureau of Economic Research.

Greene, J. D. 2013. *Moral Tribes: Emotion, Reason, and the Gap Between Us and Them*. New York: Penguin.
Hagerty, B. B. 2010. Nun Excommunicated for Allowing Abortion. NPR News, May 19. http://www.npr.org/templates/story/story.php?storyId=126985072.
Hallpike, C. R. 2004. *The Evolution of Moral Understanding*. London: Prometheus Research Group.
Hannikainen, I.R., J. Suárez, L. Espericueta, M. Menéndez-Ferreras, and D. Rodríguez-Arias. 2023. *Legal Provisions on Medical Aid in Dying Encode Moral Intuition*. Preprint. https://www.researchgate.net/publication/376480431.
Hare, R. M. 1952. *The Language of Morals*. Oxford: Oxford University Press (Clarendon Press).
Hare, R. M. 1963. *Freedom and Reason*. Oxford: Oxford University Press (Clarendon Press).
Hare, R. M. 1981. *Moral Thinking: Its Levels, Method and Point*. Oxford: Oxford University Press (Clarendon Press).
Harsanyi, J. C. 1955. Cardinal Welfare, Individualistic Ethics, and Interpersonal Comparisons of Utility. *Journal of Political Economy* 63 (4): 309–321.
Harsanyi, J. C. 1975. Can the Maximin Principle Serve as a Basis of Morality? A Critique of John Rawls's Theory. *American Political Science Review* 69 (2): 690–705.
Huang, K., J. D. Greene, and M. Bazerman. 2019. Veil-of-Ignorance Reasoning Favors the Greater Good. *Proceedings of the National Academy of Sciences* 116 (48): 23989–23995.
Huang, K., R. M. Bernhard, N. Barak-Corren, M. H. Bazerman, and J. D. Greene. 2021. Veil-of-Ignorance Reasoning Mitigates Self-Serving Bias in Resource Allocation During the COVID-19 Crisis. *Judgment and Decision Making* 16 (1): 1–19.
Kohlberg, L. 1963. The Development of Children's Orientations Toward a Moral Order. I. Sequence in the Development of Human Thought. *Vita Humana* 6:11–33.
Koop, G. J. 2013. An Assessment of the Temporal Dynamics of Moral Decisions. *Judgment and Decision Making* 8:527–539.
Light, N., P. M. Fernbach, N. Rabb, M. V. Geana, and S. A. Sloman. 2022. Knowledge Overconfidence is Associated With Anti-Consensus Views on Controversial Scientific Issues. *Science Advances* 8 (29): eabo0038.
McIntyre, Alison. 2023. Doctrine of Double Effect. In *The Stanford Encyclopedia of Philosophy (Fall 2023 Edition)*, edited by Edward N. Zalta and Uri Nodelman. https://plato.stanford.edu/archives/fall2023/entries/double-effect/.
Murphy-Berman, V., J. J. Berman, P. Singh, A. Pachuari, and P. Kumar. 1984. Factors Affecting Allocation to Needy and Meritorious Recipients: A Cross-Cultural Comparison. *Journal of Personality and Social Psychology* 46:1267–1272.
National Commission for the Protection of Human Subjects of Biomedical and Behavioral Research. 1979. *The Belmont Report: Ethical Principles and Guidelines for the Protection of Human Subjects of Research*. Available at: http://www.hhs.gov/ohrp/regulations-and-policy/belmont-report.
Patil, I., M. M. Zucchelli, W. Kool, S. Campbell, F. Fornasier, M. Calò, et al. 2021. Reasoning Supports Utilitarian Resolutions to Moral Dilemmas Across Diverse Measures. *Journal of Personality and Social Psychology* 120 (2): 443–460.
Piazza, J., and J. F. Landy. 2013. "Lean Not On Your Own Understanding": Belief That Morality is Founded on Divine Authority and Non-Utilitarian Moral Judgments. *Judgment and Decision Making* 8:639–661.
Piazza, J., and P. Sousa. 2013. Religiosity, Political Orientation, and Consequentialist Moral Thinking. *Social Psychological and Personality Science* 5 (3): 334–342.
Rawls, J. 1971. *A Theory of Justice*. Cambridge, MA: Harvard University Press.
Ritov, I., and J. Baron. 1999. Protected Values and Omission Bias. *Organizational Behavior and Human Decision Processes* 79:79–94.
Roser, D., S. Riedener, and M. Huppenbauer. 2022. *Effective Altruism and Religion: Synergies, Tensions, Dialogue*. Vol. 23. Nomos.
Shynkaruk, J. M., and V. A. Thompson. 2006. Confidence and Accuracy in Deductive Reasoning. *Memory and Cognition* 34 (3): 619–632.

Stanovich, K. E., and R. F. West. 1997. Reasoning Independently of Prior Belief and Individual Differences in Actively Open-Minded Thinking. *Journal of Educational Psychology* 89 (2): 342–357.

Strunk, W., Jr., and E. B. White. 1979. *The Elements of Style*. 3rd ed. New York: Macmillan.

Ubel, P. A. 2000. *Pricing Life: Why It's Time for Health Care Rationing*. Cambridge, MA: MIT Press.

Ubel, P. A., M. L. DeKay, J. Baron, and D. A. Asch. 1996. Cost Effectiveness Analysis in a Setting of Budget Constraints: Is It Equitable? *New England Journal of Medicine* 334:1174–1177.

Ubel, P. A., J. Baron, and D. A. Asch. 2001. Preference for Equity as a Framing Effect. *Medical Decision Making* 21:180–189.

Chapter 3
Towards a Pluralistic and Public Bioethics: Interreligious Dialogue as a Paradigm for Understanding

Michael McCarthy

Abstract This chapter explores the necessity and potential of pluralistic and public bioethics by advocating for an interreligious dialogical approach as a paradigm for discourse. Addressing the historical tension between secular and religious contributions to bioethics, it examines the interplay of values in various contexts including patient autonomy, healthcare policy, and moral theology. Drawing on the insights of Charles Taylor, the reforms of the Second Vatican Council, and Lisa Cahill's theological framework, the chapter argues that bioethics should transcend sectarian divides and adopt a dialogical methodology that fosters mutual understanding and enriches public deliberation. By engaging pluralistic perspectives, including global and non-Western approaches, the chapter aims to demonstrate how secular and religious thinkers can collaboratively address complex ethical challenges. It calls for a style of discourse that emphasizes listening, learning, and participation, in order to create a platform for addressing both shared concerns and irreconcilable differences within bioethics' global scope.

3.1 Introduction

Many of the issues addressed in this book focus on clinical questions that center on the autonomy of patients and the role of medicine. Abortion, conscientious objection, MAID all fall clearly in this tension between the patient and the healthcare provider. Bioethics, in many ways, developed at the intersection of a variety of questions pertaining to clinical practice, scientific research and were debated through the lenses of philosophy and theology—in particular a Christian theology. In the past, the Christian position may have been given a preferential priority in these debates, in the West that has become less true. However, insofar as bioethics

M. McCarthy (✉)
Neiswanger Institute for Bioethics & Healthcare Leadership, Loyola University Chicago, Maywood, IL, USA
e-mail: mpmccarthy@luc.edu

K. R. Smith (ed.), *Bioethics and Religion*, The International Library of Bioethics 116, https://doi.org/10.1007/978-3-032-33538-8_3

functions as a global practice, religious perspectives continue to hold an important point of cultural influence and still allow insights into the reasons behind certain public positions. Rather than limiting religious perspectives in public bioethical discourse, religious and secular approaches to bioethics should adopt a style of discourse within the public sphere that draws from interreligious dialogue in the spirit of pursuing a deeper understanding about the values in both secular and religious approach to bioethics.

In interreligious dialogue, differing perspectives—many of whom are convinced at the outset of the wrongness of the other's convictions—engage in conversation about the perspectives, beliefs, and necessary actions based on their religious beliefs.[1] The goal of dialogue is not to convert the other, but to better understand the other and perhaps oneself. It is in the spirit of public dialogue about pursuing the good within bioethics that Christian bioethicists can continue to make contributions to the field with their secular colleagues. This chapter argues that religious perspectives remain necessary in bioethics and the style in which they approach public discourse should draw on an interreligious approach to discourse to allow for a deeper understanding on seemingly intractable positions.

The chapter develops in three parts. Part one describes the secular by drawing first on the work of Charles Taylor, and secondly, how the Second Vatican Council shifted the Catholic Church's engagement in an increasingly secular world through moral theology and interreligious dialogue. Part two describes the place of interreligious dialogue and how that practice serves as paradigm for bridging a secular and religious divide. Part three draws on the work of Lisa Cahill's *Theological Bioethics* as a framework for a religious contribution to the public discourse of bioethics and offers several examples of what a Catholic contribution utilizing that framework could look like.[2] While some question the necessity of a discourse between secular and religious approaches to bioethics, the style in which the discourse takes place proves fundamental in determining whether a deeper understanding of each other remains possible.

---

[1] Scott Steinkerchner and SJ Clooney Francis X., *Beyond Agreement: Interreligious Dialogue amid Persistent Differences* (Blue Ridge Summit, United States: Rowman & Littlefield Publishers, Incorporated, 2010).

[2] Lisa Sowle Cahill, *Theological Bioethics: Participation, Justice, and Change* (Washington, D.C: Georgetown University Press, 2005); M. Therese Lysaught and Michael P. McCarthy editors, *Catholic Bioethics and Social Justice: The Praxis of US Health Care in a Globalized World* (Collegeville, MN: Liturgical Press Academic, 2018).

3.2 Style Matters

Secularity is marked by a certain style. Charles Taylor defines the secular age as an evolution that began in full force with the Enlightenment that saw the rise of humanism, science, and a demystification of a worldview that needed God to explain it.[3] Rather than a complete rejection of anything religious, the Enlightenment's secular approach replaced the idea of an all-powerful and controlling God at the center with the capacity for human beings to choose a path that improved humanity through objective truths proved through science and research. Taylor describes the shift in focus from a transcendent worldview with infinite possibilities to an immanent frame in which "civilized values develop, and a single-minded focus on the human good, aided by the fuller and fuller use of scientific reason, permits the greatest flourishing possible of human beings. Religion not only menaces these goals with its fanaticism, but it also undercuts reason, which comes to be seen as rigorously requiring scientific materialism."[4] Taylor observes that while a religious pursuit of the good was needed as a justification prior to the Enlightenment, in a secular age a religious perspective is less necessary, and even incompatible, with the rational and objective perspectives of science. Thus, the style of secular discourse has been to limit or reject perspectives that come from a religious perspective. The religious perspective that was most influential and rejecting of many of the features of modernity, and the subsequent secular age was the Catholic Church, perhaps no better captured than by Pope Pius X's 1907 encyclical against modernity.[5] However, Jesuit Historian John O'Malley highlights the shift that takes place at the Second Vatican Council in the 1960s, and the updating of Catholic teaching in an effort to become a more effective presence in a secular age.[6]

Until the Second Vatican Council, the Catholic Church focused on its own internal reforms and less on the challenges facing the outside world. In many ways, it resembled the view of religion that the Enlightenment rejected. However, the style of the Second Vatican Council, as O'Malley notes, marks a shift in the Catholic church's language: "In its vocabulary the style promoted a change in mind-set and in the modus operandi of the church, as from commands to invitations, from laws to ideals, from threats to persuasion, from coercion to conscience, from fault finding to

---

[3] Charles Taylor, *A Secular Age*, (Cambridge, MA: Belknap Press of Harvard University Press, 2007), 543.

[4] Taylor, 548.

[5] Pius X, "Pascendi Dominici Gregis," (September 8, 1907) | accessed December 29, 2023, https://www.vatican.va/content/pius-x/en/encyclicals/documents/hf_p-x_enc_19070908_pascendi-dominici-gregis.html.

[6] John W O'Malley, "Reform, Historical Consciousness, and Vatican II's Aggiornamento," *Theological Studies* 32, no. 4 (December 1971): 573–601.

a search for a common ground."[7] The goal of the council was to have both the capacity for internal reform and to engage an increasingly secular world as a dialogue partner. For O'Malley, the style of the Second Vatican Council proved as important as the substance of reforms that would emerge from it.

The Second Vatican Council marked an important shift in style in many areas, but two areas of relevance for this discussion: Catholic moral theology and interreligious dialogue. Richard McCormick, a Catholic priest, moral theologian, and a leading U.S. bioethicist in the U.S. after the Council, describes the tensions within moral theology between those that believed that the Church had abandoned its "classical consciousness" for a more "historical consciousness".[8] Theologian Bernard Lonergan describes the differences as one of "horizon, of total mentality". A classicist tends to have a fixed an unchanging view of human nature and the laws that govern it, while an historicist adopts a view in which "intentionality, meaning…is not fixed, static, immutable, but shifting, developing going astray, capable of redemption."[9] Thus, interpretation, particularly in matters of moral teachings, after the council resulted in polarized responses of how best to engage the historical context in which the Catholic Church existed. McCormick noted several areas that prompted a reexamination after Vatican II. For bioethics, three areas that continue to require attention is the "social character of the moral life"; "the centrality of the person"; and "the tentativeness of moral formulations."[10] In each of these areas, greater attention is paid to the actions of individuals, the relational character of moral actions, and the importance of the contexts that shape one's actions. This attention to history and the contexts in which human persons find themselves, also contributed to the rethinking of how the Catholic Church engaged other religions.

Nostra Aetate, the declaration on the relation of the Catholic Church to non-Christian religions, emphasizes the unity among all people and calls for a genuine dialogue with other religions to "preserve and promote good things…as well as socio-cultural values…"[11]

---

[7] John W O'Malley, "'The Hermeneutic of Reform': A Historical Analysis," *Theological Studies* 73, no. 3 (2012): 544.

[8] Richard A. McCormick, *The Critical Calling: Reflections on Moral Dilemmas since Vatican II*, Moral Traditions Series (Washington, D.C: Georgetown University Press, 2006), 9. The distinction between classicist and historicist worldview became a key way of describing the differences in the reception of teachings after Vatican II.

[9] Bernard Lonergan, "The Transition from a Classicist Worldview to Historical Mindedness," in *A Second Collection*, vol. 13, Robert M. Doran and John D. Dadosky, eds., Volume 13 (University of Toronto Press, 2016), 3–10.

[10] McCormick, *The Critical Calling*, 14–17.

[11] "Nostra Aetate," accessed December 28, 2023, https://www.vatican.va/archive/hist_councils/ii_vatican_council/documents/vat-ii_decl_19651028_nostra-aetate_en.html.

The document replaced—and again speaks to McCormick's insight about the tentativeness of moral positions—the long-held teaching that "There is no salvation outside of the Church".[12] In fact, "The Catholic Church rejects nothing that is true and holy in these religions. She regards with sincere reverence those ways of conduct and of life, those precepts and teachings which, though differing in many aspects from the ones she holds and sets forth, nonetheless often reflect a ray of that Truth which enlightens all..."[13] While the document does not say that the position of other religions are equal to the Catholic position or the teachings of other religions ought to be followed, it does open the possibility of dialogue on questions, beliefs, and practices that differ from those of the Catholic Church. Like the developments within moral theology, there emerges a curiosity and encouragement to enter into dialogue with those who hold different perspectives.

The impact of the Second Vatican Council, and the interpretation argued for here, is not without its critics. Engelhardt saw the Second Vatican Council as sowing chaos both into the Roman Catholic and mainline Protestant Churches. He notes "The winds of aggiornamento had given Roman Catholics and mainline Protestant's confidence in the view that they should be able, without invoking particular, sectarian Christian commitments, to speak to the secular world and even to speak in the secular world's terms while retaining their own integrity."[14] He finds that the ripple-effect of Vatican II created space within bioethics to develop compromise positions that became devoid of any perceived Christian character. This new context gave rise to a Christian bioethics that was "nothing more than secular bioethics done by bioethicists who claimed to be Christian."[15] However, Engelhardt's interpretation maintains a sectarian divide between Christians and non-Christians, Christian ethics and secular ethics, a divide that the Council sought to overcome or at least engage. The goals of the Second Vatican Council are better reflected in McCormick's more historically conscious approach to moral theology and Nostra Aetate's openness to dialogue with other religions. An historically consciousness approach to the public practice of bioethics aims to engage secular and religious thinkers in an honest dialogue about both what to do when faced with a particular problem and to move toward thicker conversations about the significance of pursuing one action instead of another.[16]

---

[12] Michael D. Driessen, "Catholicism and Interreligious Dialogue," in *The Global Politics of Interreligious Dialogue: Religious Change, Citizenship, and Solidarity in the Middle East*, ed. Michael D. Driessen (Oxford University Press, 2023), 91.

[13] "Nostra Aetate."

[14] H. T. Engelhardt, "The Recent History of Christian Bioethics Critically Reassessed," *Christian Bioethics* 20, no. 2 (2014): 151.

[15] Engelhardt, 150.

[16] John H. Evans, *Playing God?: Human Genetic Engineering and the Rationalization of Public Bioethical Debate*, Morality and Society (Chicago: University of Chicago Press, 2002), 4–7.

3.3 Bioethical Discourse in a Pluralistic Key

Sociologist John Evans divides bioethics into four jurisdictions (research, healthcare ethics consultation, public policy, and cultural bioethics), and sees a role for theological bioethics engaging in questions about ends that fall mainly within public policy and cultural bioethics questions.[17] He notes that in the areas of research ethics and healthcare ethics consultation there is very little space for debates about ends. In research ethics, the function of bioethicists is applying legally codified principles (respect for persons, beneficence, and justice) in approving institutionally based research.[18] For healthcare ethics consultants, while principles play a primary role, in most situations there is a heavy emphasis placed on respecting the autonomy of the patient's choices as the primary goal.

Insofar as theology is engaged in a discussion of ends, there is space within public policy and cultural bioethics debates to explore the values that shape more public facing bioethical questions.[19] Within public policy, one can look to the role theologians have played on the U.S. Presidents Commissions on Bioethics or theologians within the EU who sit on various committees on research and new technologies.[20] The role of theologians in these debates center both on particular actions that should be taken on an issue and in describing what values or ends are being pursued by choosing one action over another. Evans argues that "theology is one of the sources of non-instrumental and non-individualistic values in contemporary capitalist societies, and how a large percentage of the public thinks, so theological input is very important."[21] While it need not set the standard or direction a discourse should be taken, theologians can push for a deeper understanding of the rationale behind both secular and religious positions. Communicating across traditions, as seen from the effects of Vatican II, is not a new phenomenon in theological discourse. Rather than approaching a dialogue as moral strangers, interreligious dialogue provides a paradigm for theological bioethicists to engage in a pluralistic discourse that can be helpful in two ways.

First, interreligious dialogue approaches the discourse to move beyond tolerance of the other and towards a mutual appreciation and recognition of the other perspective but not with the goal of conversion. Theologian Roger Haight argues, "Religious pluralism is necessary because we live in history. Historical

---

[17] John H. Evans, "The Social Context of Religion in the Jurisdictions of Bioethics," *The American Journal of Bioethics* 20, no. 12 (December 1, 2020): 1–4; John H Evans, *The History and Future of Bioethics: A Sociological View* (New York: Oxford University Press, 2012).

[18] "Federal Policy for the Protection of Human Subjects" ('Common Rule," Text, HHS.gov, June 23, 2009, https://www.hhs.gov/ohrp/regulations-and-policy/regulations/common-rule/index.html.

[19] Evans, "The Social Context of Religion in the Jurisdictions of Bioethics," 2.

[20] Lisa Cahill, "Public Theology and Bioethics," in Sebastian Kim and Katie Day, editors, *A Companion to Public Theology* (Boston, MA: BRILL, 2017), 369–89; Cathriona Russell, Linda Hogan, and Maureen Junker-Kenny, eds., *Ethics for Graduate Researchers: A Cross-Disciplinary Approach* (London: Elsevier, 2013).

[21] Evans, "The Social Context of Religion in the Jurisdictions of Bioethics," 2.

consciousness entails the fact that all experience is local and particular religious experience not excepted."[22] Therefore, to accept an argument simply because it is, or is not, from a religious perspective cannot merit serious attention. A religious and a secular approach within a public and pluralistic bioethics necessarily takes into account the historical, local, and particular cultural context in which the discourse is taking place. A pluralistic approach to bioethics acknowledges the importance of historical contexts and experiences that give shape to the questions of bioethics and the various secular and religious approaches that attempt to give answers to these questions.

Second, a pluralistic approach is rooted in a dialogical framework that seeks to establish common perspectives and hoped for ends among interreligious interlocutors. Haight again offers an important insight on what constitutes dialogue more broadly. He describes dialogue as involving "a give and take, a communication that requires not just testifying, announcing, and teaching but also listening, hearing, and learning. It presumes mutual recognition and at some level a sense of equality among the participants as a characteristic of the conversation. Dialogue is much more than actual exchange. It also defines a climate, a fundamental institutional attitude, and a strategy toward the other institution."[23] Of equal importance, "Dialogue also rules out any proselytism or aggressive, conquering evangelization...[it] must respect the freedom of the dialogue partners."[24] Thus, if interreligious dialogue is to provide insights from secular and religious perspectives, there needs to be a genuine openness to the position of the other in the exchanging of ideas. Moreover, these pluralistic discourses must respond to the realities in which they are confronted, not in defense of some intrinsic position. A pluralistic approach would revive the place that humanistic values hold and how those values can establish a framework through which further dialogue can be pursued.

There is perhaps no better example of a pluralistic bioethical discourse drawing from philosophy, theology, law, and science than the 1974 National Commission for the Protection of Human Subjects of Biomedical and Behavioral Research. The questions taken up by the Commission originated in response to unethical practices in research, the need to protect human subjects, and to establish guidelines for the responsible conduct of research. Albert Jonsen recalls that two essays that had a significant impact on the principles chosen by the commission, an essay by H. Tristam Engelhardt and the other by Thomas Beauchamp.[25] The Commission discussed the merits of each essays approach, Engelhardt's focusing on respect for persons and a focus on the best interest of human subjects and Beauchamp on the concept of distributive justice. While the language of each was philosophical in

[22] Roger Haight, *The Nature of Theology: Challenges, Frameworks, Basic Beliefs* (Maryknoll, NY: Orbis Books, 2022), 51.

[23] Roger Haight, *The Future of Christology* (New York, NY: Bloomsbury Publishing, 2007), 134.

[24] Haight, 134.

[25] Albert R. Jonsen, "On the Origins and Future of the Belmont Report," in ed. James Childress, Eric M. Meslin, and Harold T. Shapiro, *Belmont Revisited: Ethical Principle for Research with Human Subjects*, (Washington, D.C: Georgetown University Press, 2005), 4.

nature, the principles reflected values that had deep roots in the religious tradition of those on the Commission. The Commission still serves an example pluralistic and public bioethical discourse engaging both secular and religious thinkers.

If a pluralistic discourse is to be possible in bioethics, it is important that to acknowledge that there will be positions on which certain secular and religious thinkers will not agree. Nevertheless, Charles Camosy draws on the thought of Alasdair MacIntyre to argue that even if there exists foundational incompatibilities between traditions, "such exchanges, if they be about something other than power, would have to be normed by certain practices that create the conditions for the possibility of debate: nonviolence, truth-telling, sociopolitical equality and so-on."[26] In both instances, the discourse between secular and religious thinkers has to center on the topic being discussed, knowing that end positions on specific cases may not yield a desired result as entrenched as the Belmont Principles. Nevertheless, pluralistic bioethical discourse can reveal deeper insights into rationale and establish common ground on which to address real challenges facing bioethics.

3.4 Pluralistic Bioethics for a Global World

Establishing a shared framework through which a pluralistic discourse can take place within bioethics requires pursuing ends that both secular and religious thinkers find worthwhile, with an openness to listening, hearing, and learning more in conversation. Despite the critique of some Christian bioethicists, Lisa Cahill's framework of participation, justice, and change provides a starting point for a pluralistic approach to bioethics.[27] While finding critical reception within more sectarian approaches to Christian bioethics, her framework has drawn more positive engagement from secular thinkers.[28]

Lisa Cahill's 2005 *Theological Bioethics: Participation, Justice, Change*, provides a necessary framework for how secular and religious thinkers can approach public discourse within bioethics. Cahill's argument shifts from a more polarizing perspective of a secular versus religious approach, opting instead for an understanding of bioethics as a public discourse.[29] There are three strengths in framing bioethics as a public discourse, rather than either a secular or religious practice. First, this approach aligns with Evan's sociological understanding of bioethics and falls neatly

[26] Charles C Camosy, "Irreligion, Alfie Evans, and the Future of Bioethics," *The Journal of Medicine and Philosophy: A Forum for Bioethics and Philosophy of Medicine* 46, no. 2 (April 1, 2021): 166.

[27] Cahill, *Theological Bioethics: Participation, Justice, and Change*; Engelhardt, "The Recent History of Christian Bioethics Critically Reassessed," 154.

[28] Russell Blackford and Udo Schüklenk, "Religion at Work in Bioethics and Biopolicy: Christian Bioethicists, Secular Language, Suspicious Orthodoxy," *The Journal of Medicine and Philosophy: A Forum for Bioethics and Philosophy of Medicine* 46, no. 2 (April 1, 2021): 176–77.

[29] Cahill, *Theological Bioethics: Participation, Justice, and Change*, 20–21.

within the jurisdiction of public policy and cultural bioethics. Secondly, public bioethics requires dialogue and engagement. Drawing on the importance of dialogue emphasized by Haight for religious pluralism, dialogue should avoid proselytizing or a goal of evangelizing opting instead to listen, hear, and learn. Finally, from this dialogue emerges a response that engages public practice requiring action. Rather than dismissing a secular or religious approach, a public approach draws on a dialogue between differing methodologies to address the issues that confront bioethics today and need a response.

Identifying a common starting place, or a common inequality that needs to change, can provide a helpful place from which to begin a pluralistic discourse for bioethics.[30] In 2018, a collection of authors focused on a range of bioethical issues facing U.S. healthcare systems and draws on Catholic Social Teaching as a resource for addressing these challenges. The authors focused on a variety of questions including: gun violence, climate change, racism and the social determinants of health, among others.[31] The unique aspect of the topics chosen is that they are typically not addressed as theological issues and pertain more to the domain of secular bioethics. However, by drawing on Catholic Social tradition they argue, for example, that the reality of gun violence in the U.S. requires a greater commitment to implementing Hospital Violence Intervention Programs (HVIPs) "to interrupt the epidemic of gun violence."[32] Another author describes the health disparities that Black American's face with respect to end-of-life care and the need to establish solidaristic practices within the local community to "address the lingering effects of medical mistrust."[33] Within the U.S. many of injustices captured fail to be addressed in most mainstream Catholic bioethical discourse, but are much more readily addressed within a secular approach to bioethics.[34] Providing that common starting place, can even be more easily established when looking beyond the western framework for bioethics and towards a global approach.

The debate between secular and religious approaches to bioethics is one that is particular to the Western origins of bioethics. Rather than beginning from a religious or secular methodology, global bioethics scholars tend to begin by addressing

---

[30] Michael McCarthy, Mary Homan, and Michael Rozier, "There's No Harm in Talking: Re-Establishing the Relationship Between Theological and Secular Bioethics," *American Journal of Bioethics* 20, no. 12 (2020): 5–13.

[31] M. Therese Lysaught editor and Michael P. McCarthy editor, *Catholic Bioethics and Social Justice*.

[32] Michelle Byrne, et al., "Healthcare Providers on the Frontline: Responding to the Gun Violence Epidemic," in M. Therese Lysaught and Michael P. McCarthy, editors, *Catholic Bioethics and Social Justice: The Praxis of US Healthcare in a Globalized World*, 43.

[33] Sheri Bartlett Browne and Christian Cintron, "Racial Disparities at the End of Life and the Catholic Social Tradition," in M. Therese Lysaught and Michael P. McCarthy, editors, *Catholic Bioethics and Social Justice: The Praxis of US Healthcare in a Globalized World*, 158.

[34] Alan Elbaum, "Black Lives in a Pandemic: Implications of Systemic Injustice for End-of-Life Care," *The Hastings Center Report* 50, no. 3 (2020): 58–60; Sean Philpott-Jones, "Mass Shootings, Mental Illness, and Gun Control," *The Hastings Center Report* 48, no. 2 (2018): 7–9.

the historical issues that affect the communities in which they live and work.[35] One of the earliest examples of highlighting these different methodological approaches emerges in an edited volume that explores an alternative starting point for bioethical discourse. These authors explore a different approach to bioethics from a Filipino context that emphasizes the way in which religion, culture, and ethical questions facing their community engage one another and are not separated easily into secular or religious approaches as they are in a Western approach.[36] Jacquineau Azetsop incorporates the environmental, political, and public health challenges in an African context. He draws liberally on theological and secular bioethical insights to emphasize the complex bioethical issues from his context as an African social ethicist.[37] Finally, Alexandre Martins work, a Brazilian theologian, takes an explicitly theological approach that begins with the experience of those suffering because of their socio-economic marginalization and how a liberationist approach begins "from below" in order that "the poor become agents of their own liberation."[38] Like the authors in Tan-Alora and Lumitao's volume from the Philippines and Azetsop from Africa, Martins's insights from a Brazilian context demonstrate how a non-western approach to bioethics functions as a public discourse that does not discriminate between religious or secular in search for answers to historical challenges. Their approaches point to the possibilities of engaging in bioethics as a public discourse between secular and religious thinkers instead of a polemical one.

3.5 Conclusion

In a certain sense, the question of whether a religious approach is an acceptable one depends on who is asking the question.[39] Bioethics is, and has always been, an interdisciplinary and public discourse that draws on a variety of methodological sources

[35] Alexandre Martins, *The Cry of the Poor: Liberation Ethics and Justice in Health Care* (Lanham, Md.: Lexington Books, 2019); Jacquineau Azetsop, ed., *Integral Human Development: Challenges to Sustainability and Democracy* (Eugene, Oregon: Pickwick Publications, 2019); Tan-Alora Angeles., Lumitao, Josephine M., *Beyond a Western Bioethics : Voices from the Developing World* (Washington, D.C.: Georgetown University Press, 2001).

[36] Tan-Alora, *Beyond a Western Bioethics: Voices from the Developing World.*

[37] Azetsop, *Integral Human Development*; Jacquineau Azetsop, "Challenges in Global Health, Culture, and Ethics in Africa," *Journal of Moral Theology* CTEWC Book Series, no. 1 (2021): 119–30.

[38] Martins, *The Cry of the Poor: Liberation Ethics and Justice in Health Care*, xiv.

[39] Critics of whether this discourse is necessary, let alone desired, exist on both the religious and secular divide. See: Timothy F. Murphy, "In Defense of Irreligious Bioethics," *The American Journal of Bioethics* 12, no. 12 (December 1, 2012): 3–10; Udo Schuklenk, "On the Role of Religion in Articles This Journal Seeks to Publish," *Developing World Bioethics* 18, no. 3 (September 2018): 207–207; Engelhardt, "The Recent History of Christian Bioethics Critically Reassessed"; Stanley Hauerwas, "How Christian Ethics Became Medical Ethics: The Case of Paul Ramsey," *Christian Bioethics: Non-Ecumenical Studies in Medical Morality* 1, no. 1 (March 1, 1995): 11–28.

to address what are often complex and multifaceted issues. Nevertheless, there are religious and secular bioethicists who still believe that a more sectarian approach is warranted. This essay argues the opposite, that in fact, Bioethics is richer as a pluralistic and public practice.

Whether the discourse of bioethics moves forward in a pluralistic manner, rests in the style with which one approaches it. From a Christian perspective, the spirit of Vatican II and its attention to history recasts the way in which Catholic thinkers engaged moral theology and created an openness to dialogue with those with whom they knowingly had disagreements. Interreligious dialogue does not aim to convert a Buddhist to a Catholic, any more than a religious thinker aims to convert a secular thinker. Yet, the practice of pluralistic dialogue allows for the opportunity to come to a deeper understanding of the other and oneself. Within bioethics, it creates a possibility of identifying complementary approaches on which to address the complex bioethical challenges facing the world today but will inevitably identify intractable issues on which compromise is not possible. Nevertheless, the problems facing bioethics in its global reality needs complex thinking from a variety of approaches in order to confront the many questions that reach beyond a secular and religious divide.

Chapter 4
Conscience Beyond Religion: Exploring Secular Conscientious Objection and Provision

Abram Brummett

Abstract This chapter examines the scope of conscientious objection in healthcare, proposing that protections be extended to include secular conscientious objections and provisions founded on moral principles rather than solely religious beliefs. Traditionally, conscientious objection has permitted clinicians to refuse participation in specific medical services on religious grounds; however, this chapter argues that secular clinicians should have similar protections when motivated by deeply held moral convictions. The first section considers secular objections to the treatment of neurologically devastated patients, focusing on ethical and legal concerns, including interpretations of the Americans with Disabilities Act concerning patients in persistent vegetative states or permanent unconsciousness. The second section addresses the legal disparity between conscientious objection and conscientious provision, particularly within Catholic healthcare institutions where secular clinicians may wish to offer services that are institutionally restricted. The chapter proposes limited protections for conscientious provisions, such as referrals or prescriptions, which require minimal institutional resources. By advocating for a broader framework for conscience rights, this chapter argues for the need for healthcare systems to respect all moral convictions, whether religious or secular.

4.1 Introduction

The debate over the role of clinician conscience in health care is a fault line between religious and secular camps in bioethics. A clinician invokes a conscientious objection when they refuse to provide a legal and professionally accepted medical good or service on the grounds that doing so would violate their core religious or moral beliefs (Wicclair 2011). Legal protections for individuals and institutions that

A. Brummett (✉)
Department of Foundational Medical Studies, Oakland University William Beaumont School of Medicine, Rochester, MI, USA
e-mail: abrummett@oakland.edu

K. R. Smith (ed.), *Bioethics and Religion*, The International Library of Bioethics 116, https://doi.org/10.1007/978-3-032-33538-8_4

conscientiously object to performing, or assisting in the performance of, abortion or sterilization procedures have been in place since the passage of the Church Amendments on the heels of the Roe v. Wade decision in 1973. Subsequent federal and state conscience clauses have been passed, resulting in robust protections for conscientious objection that cover virtually any conscientious objection unless it either 1-violates federal or state antidiscrimination statutes, or 2-is invoked when a patient shows up to the emergency department in need of urgent care. Providers are not even legally required to disclose that their refusals are based on moral (rather than medical) grounds or to refer patients to willing providers (Fox 2022).

Despite the inclusion of religious *or moral* values in most conscience clauses, conscientious objection has primarily been utilized by clinicians with conservative religious commitments who refuse to provide medical treatments for abortion, contraception, sterilization, or physician-assisted suicide (Rhodes 2019; Dickens 2014). Much of the bioethics literature over the last 50 years has tended to focus on defending (Hershenov 2021), rejecting (Schuklenk 2019), or constraining (Card 2020) these forms of conscientious objection. This essay explores the possibility of secular conscientious objection and provision beyond religion, taking seriously the inclusion of "moral values" as a defensible basis for a conscience claim. The term "secular" is used heuristically to refer to clinicians whose moral values generally oppose "conservative," often "religious," "right to life" views (though it is important to acknowledge that many "conservative" or "religious" clinicians may also be compelled by the conscience claims explored in this essay).

Part one of this essay focuses on conscientious objections to providing care for neurologically devastated patients (e.g., vegetative or brain dead patients). It is argued that these conscientious objections ought to be legally permitted, which will require excluding permanent unconsciousness as a disability to be protected by the Americans with Disabilities Act. Part two focuses on conscientious provision in the context of clinicians who wish to provide a legal and professionally accepted medical good or service that is prohibited by the Catholic hospital that employs them. Catholic hospitals operate in line with the Ethical and Religious Directives for Catholic Health Care Services (ERDs), which prohibits medical treatment for the purpose of contraception, sterilization, abortion, and physician-assisted-suicide. Conscientious provisions are not currently protected by U.S. law, but this section will argue that some ought to be.

The arguments for secular conscientious objection explored in this essay presume a "reasonability view" of conscience (Card 2020; Eberl 2019a; Brummett and Eberl 2025). The reasonability view does not permit any conscience claim, but only those that satisfy the reciprocity criterion of public reason, which asks interlocutors to offer one another reasons they can reasonably expect persons committed to different comprehensive doctrines to accept (Rawls 1972). For example, it is possible to have reasonable disagreement over the practice of physician-assisted-suicide using only arguments cast in public reason. Those in favor may claim that a reasonable goal of medicine is to provide patients with a peaceful death (Miller and Brody 1995), while those opposed may claim that medicine ought to be focused solely on the healing of broken bodies (Callahan 1992). Importantly, the reasonability view prohibits conscience claims that cannot be given a plausible public reason defense. For example, conscience claims based on invidious discrimination, such as the refusal to dispense PrEP (Pre-Exposure Prophylaxis)[1] only to gay or bisexual men so as not to "enable immoral sexual behavior" would be prohibited (Brummett 2020a).[2] Similarly, conscientious objections that rely on purely faith-based premises (such as a Jehovah's Witness clinician who refuses to administer blood products)[3] would also be prohibited by the reasonability view.

4.2 Part One: Secular Conscientious Objection

Some clinicians experience significant moral distress when families demand continued treatment of neurologically devastated patients. For the purposes of this essay, "neurologically devastated" patients are those who are, and virtually certain to remain, unconscious. Specifically, neurologically devastated patients include the permanently vegetative or those who have been pronounced dead by neurological criteria. Here, the possibility of individual and institutional conscientious objection to treating this patient population is explored.

In the U.S. alone, there are estimated to be between 14,000 and 35,000 vegetative patients (Ashwal et al. 1994) estimated to cost the health system between 1 and 7 billion dollars a year (Schneiderman and Jecker 2011, 47). When considering this patient population, it is critical to distinguish between vegetative patients that retain a significant chance of recovering consciousness from those that are virtually certain to remain unconscious. Recent evidence suggests that clinical indications, such as age and the nature and extent of injury, give some vegetative patients a greater chance of recovery than was previously believed (Giacino et al. 2018). Additionally, novel therapies such as deep brain stimulation, pharmacology, and electrophysiology have proven to aid in the recovery process for these patients (Schiff et al. 2009; Edlow et al. 2021). One widely cited study demonstrated that some patients diagnosed as vegetative using bedside assessment actually retained "covert consciousness" when fMRI brain imagining shockingly revealed the presence of brain function similar to healthy controls (Monti et al. 2010). Alternatively, other vegetative patients do not have a significant chance of developing or recovering consciousness. For some, a congenital malformation, such as anencephaly, has no corrective therapy because there is no higher brain matter upon which to intervene. For others,

---

[1] PrEP is a safe and effective daily medication that works to block important pathways used by HIV to set up infection.

[2] Recent legal actions in the U.S. have attempted to empower clinicians and insurance providers who wish to refuse PrEP to patients on the basis of their sexual orientation, arguing that compelling such treatment would violate the Religious Freedom Restoration Act (Scherer 2022).

[3] Although it is difficult to find documented examples in the academic literature, a Google search for the phrase "can a Jehovah's Witness nurse administer blood" will reveal active discussion boards of Jehovah's Witness nurses debating the issue.

the nature of the injury (e.g., a severe anoxic injury) and clinical testing (e.g., somatosensory evoked potential) can point to virtual certainty of permanent unconsciousness within 3 days of injury (Zandbergen et al. 2006). These patients can linger on in a vegetative state for many years and will develop pressure sores, muscle contractures, and blood clots without regular medical care. Nevertheless, families regularly choose this fate for their loved ones.

Families have also requested continued treatment for the bodies of patients who have been pronounced dead by neurological criteria. After being pronounced brain dead, Jahi McMath's body was sustained for years in New Jersey, a state with a conscience exemption to brain death, because of her mother's demands for continued treatment (Truog 2018). Family demands for treatment of brain dead bodies can also carry significant legal weight in states without a conscience exemption to brain death. For example, in Michigan, clinicians were legally compelled to continue treating the corpse of Titus Cromer as a judge repeatedly stayed the withdraw of ventilatory support due to family objection (Shamus 2020). These cases are likely to become more widespread as some bioethicists have called for a nationwide exemption to brain death following the New Jersey example (Veatch and Ross 2016). Members of the Uniform Commission recently met to discuss adding a conscience clause to accommodate those with a philosophical or religious objection to whole brain death (Pope 2023). Such a revision to the Uniform Determination of Death Act would likely be adopted by the vast majority of states in the U.S., and while it is difficult to estimate precise numbers, experts opine that as much as 10% of the population would utilize the conscience exemption to brain death (Veatch and Ross 2016, 66). A recent study of public attitudes showed that 41.9% of respondents support adding a conscience clause to require clinicians to continue treating an individual with brain death if their family rejects brain death. (Ludka et al. 2025).

If a surrogate requests aggressive treatment for a neurologically devastated patient, a clinician may refuse, but a justification for the refusal is needed. One possible justification for unilateral withdrawing or withholding treatment over family dissent is that treatment would be physiologically futile. Physiologically futile interventions are those with no chance of achieving the desired goal (Bosslet et al. 2015). If the goal of the surrogate is to keep the patient alive, then physiological futility does not apply, because patients can be maintained with medical support for years, even decades, in both a vegetative state and after brain death has been pronounced.

Another possible justification is that treatment would be potentially inappropriate. Potentially inappropriate interventions have at least some probability of achieving the desired goal, but clinicians believe there are sufficient competing ethical considerations (e.g., clinician moral distress, stewardship of medical resources, low probability of success, lack of medical benefit) to justify withholding or withdrawing them (Bosslet et al. 2015). The medically inappropriate standard is problematic for neurologically devastated patients because professional consensus on whether treating these patients crosses a threshold of medical inappropriateness such that treatment may be withheld or withdrawn over family request has proven elusive. After all, one goal of medicine is to sustain life, and providing treatment for neurologically devastated patients is very likely to achieve this goal. Furthermore, applying the medically inappropriate language is challenging for neurologically devastated patients because it presumes notions of "benefit" and "harm" that are difficult to apply to a patient without consciousness. As Syd Johnson remarks, "It is difficult to make sense of the claim that anything can be in an unconscious individual's current "best interests" if having interests at all requires consciousness" (Johnson 2022, 33).

Further, presumptive evidence of an elusive professional consensus on the medical appropriateness of sustaining the life of vegetative patients can be found between the "five societies statement"[4] by Bosslet et al. and the subsequent statement by Kon et al. (Bosslet et al. 2015; Kon et al. 2016). In the original five societies statement, some examples of "potentially inappropriate" treatment were provided. Unsatisfied with this, one of the professional societies that co-drafted the original statement (the Society for Critical Care Medicine) subsequently published their own statement to provide a clearer definition of medically inappropriate care. What was explicitly added by the Society of Critical Care Medicine was a statement that the appropriate goals of ICU care include "a reasonable expectation for survival outside the acute care setting *with sufficient cognitive ability to perceive the benefits of treatment*" (Kon et al. 2016, emphasis added). Presumably, members from the Society of Critical Care Medicine suggested this when drafting the five societies statement, but agreement could not be reached as to whether to include this language.

Instead of "physiological futility," or the "potentially inappropriate" standard that presumes notions of benefit, harm, or quality of life that have proven difficult to apply to neurologically devastated patients, it may be better to classify these clinician refusals as acts of conscientious objection. Grounding refusals in clinician conscience is more appropriate for those who may reasonably conclude that sustaining neurologically devastated patients in a state of permanent unconsciousness is a violation of human dignity. The language of conscience and human dignity may more accurately track the real reasons for a clinician's refusal, who have described these patients as "lying in an open grave" (Crisci 1995), lacking "the most essential feature of the human being" (Rich 1997), and "better off dead" (Payne et al. 1996). A study of nurses caring for vegetative pediatric patients described the work as "more for the parents" and report significant distress caring for unconscious bodies that "smell of death" (Montagnino and Ethier 2007). As a conscience claim, such a justification would not need to resolve intractable disagreements about a patient's best interests or quality of life, but only be a reasonable and genuinely held core belief. Furthermore, grounding the refusals in conscience will be legally forceful given the strong federal and state conscience protections for clinicians.

A key objection to reframing a clinician's refusal as an act of conscientious objection is that it amounts to disability discrimination under the Americans with Disabilities Act (ADA). The question then becomes, is permanent unconsciousness

[4] The five professional societies include: the American Thoracic Society, the American Association for Critical Care Nurses, the American College of Chest Physicians, the European Society for Intensive Care Medicine, and the Society of Critical Care Medicine.

a mere disability, or something worse? Hillary Chua has argued that patients with disorders of consciousness should receive legal protection from discrimination under the ADA (Chua 2020). As such, an individual clinician would be prohibited from refusing treatment for vegetative patients in the same way they are prohibited from refusing treatment based on race or religion. Chua's claim is contested by Andrew Batavia, a quadriplegic lawyer who helped draft the ADA. Batavia argues that "there is a fundamental difference between people who are permanently unconscious and other people who meet the disability definition of the ADA…the former lack the human quality [of consciousness] that warrants their characterization as people with disabilities and the protections of the anti-discrimination laws" (Batavia 2002).

While this debate remains unsettled, the reasonability view of conscientious objection only requires that both sides have a publicly defensible rationale on the question of whether permanently unconscious patients ought to be legally protected under the ADA. As this debate continues to unfold, it will be critical to distinguish between vegetative patients with a reasonable chance of recovery from those without, and the conscience exemption should only apply to patients who are virtually certain to remain unconscious. Additionally, with the recent conscience exemption to brain death proposed for consideration by the Uniform Commission, those who defend ADA protections for vegetative patients will need to either also entertain defending ADA protections for patients who have been pronounced brain dead or explain why this absurd result does not follow from their view.

If the argument for individual conscientious objection to treating neurologically devastated patients is accepted, it is worth exploring whether *institutional* conscience claims of this sort should also be permitted (Wasserman et al. 2026). Institutional conscience protects an organization—shaped by its mission, vision, and values—to enact system-wide prohibitions on goods and services that conflict with its core values (Wildes 1997). The right to institutional conscience is already legally protected in the U.S., but it tends only to be leveraged by faith-based organizations. Most commonly, Catholic hospitals enact wholesale institutional prohibitions on medical goods and services aimed at abortive, sterilization, or contraceptive purposes (Bishops 2018). These treatments are refused by entire Catholic health systems on the grounds that they violate the view of human dignity endorsed by the Church. These claims of institutional conscientious objection for faith-based institutions have proven legally stable. No judge would think of compelling a Catholic hospital to provide treatments that violate its view of human dignity, even when those treatments are legal, within the standard of care, and requested by patients or families.

If religious institutions can enact system-wide prohibitions on medical treatment that violates core values, then secular institutions should be afforded the same right. Even bioethicists that defend the concept of institutional conscience for Catholic hospitals readily acknowledge that secular institutions can make similar claims (Bedford 2016). For example, Elliott Bedford states that "arguments in support of the concept do not see institutional conscience as a solely religious phenomenon" (Bedford 2012, 416). A secular institution may reasonably determine that indefinitely sustaining the life of neurologically devastated patients violates a core value that consciousness is a minimum requirement of human dignity.

4.3 Part Two: Secular Conscientious Provision

There is an asymmetry in the treatment of conscience in U.S. law whereby conscientious objections are protected but conscientious provisions are not (Sepper 2012). In other words, a physician is legally protected when refusing to provide a legal and professionally accepted medical good or service when employed at a secular institution that permits this service, but the reverse is not true. A physician is not legally protected should they attempt to provide a legal and professionally accepted medical good or service when employed at a Catholic institution, governed by the Ethical and Religious Directives for Catholic Health Care Services (ERD), that prohibits the service. Here, the possibility of ERD-prohibited and ERD-permitted conscientious provisions are explored.

4.3.1 ERD-Prohibited Conscientious Provision

Before arguing that some ERD-prohibited conscientious provisions deserve legal protection, it will be necessary to first consider and reject three arguments *for* the conscience asymmetry. First, some may argue that Catholic institutional conscience is a sufficient reason to uphold the conscience asymmetry. On this view, Catholic institutions are grounded in a religious tradition with a well-developed view of human dignity and other values that are used to proscribe the provision of some legal and professionally accepted medical goods and services. However, as mentioned above, there is no reason to think that institutional conscience is a concept uniquely available to religious hospitals. Like Catholic hospitals, secular hospitals operate in ways that promote and presume a wide array of moral, metaphysical, and epistemological commitments (Brummett and Eberl 2022; Brummett 2020).

Second, it may be objected that even if secular institutional conscience is granted, there is a practical difference between the positive values (e.g., to provide birth control for contraception) of secular institutions and the negative values (e.g., to not provide birth control for contraception) of religious institutions because positive institutional values can still be upheld in the face of individual clinician objection by simply having another clinician provide the medical good or service. Alternatively, if even one clinician wishes to conscientiously provide a medical good or service in violation of a negative institutional value, then the institutional value cannot be upheld (Brummett 2020b). However, it is not clear that the distinction between positive and negative institutional values can be meaningfully maintained because secular institutions with "positive institutional values" can reframe their commitments in negative terms. So, for example, a secular institution could oppose having any patients face the indignity of a conscientiously objecting clinician. Thus reframed, if even one clinician conscientiously objects to any legal and professionally accepted medical good or service, then a "negative institutional value" has been violated. Therefore, this objection cannot justify the conscience asymmetry.

A third argument to support the asymmetry asserts there is a moral difference between omission and commission such that omissions deserve legal protection while commissions do not. On this view, omissions deserve unique legal protection because negative duties of omission obligate absolutely, while people have wide discretion in responding to positive duties of commission. For example, the negative duty not to murder may appear to be a duty that binds absolutely, while the positive duty to give to charity is discretionary. If this asymmetry holds generally—if negative prohibitions bind absolutely, while positive duties do not—then *forcing* a clinician to provide a treatment when they believe they are ethically obligated not to do so may necessarily involve them in wrongdoing, while *prohibiting* a clinician from providing a treatment may not involve a violation of their moral obligations. It may seem to follow that there is less reason to legally protect conscientious provision than conscientious objection, and that the conscience asymmetry is justified. However, even if the omission/commission distinction is valid, it does not follow that it deserves the same moral consideration in the context of the doctor-patient relationship where a physician has, by virtue of their role, taken on significant positive duties to promote the health of their patient. Further, and more significantly, even if it could be established that the omission/commission distinction is valid, it does not follow that *only negative values deserve protection* (Wicclair 2009). In determining whether a conscience claim deserves protection, what matters is whether, in the absence of conscience protections, some *threshold* will be crossed when it comes to undermining a clinician's core values. Such a threshold can be crossed for commissions because clinicians who wish to provide a service to a consenting patient may feel intense shame or loss of self-respect if they are prohibited from doing so. The fact that compelled omission may be worse than compelled commission is irrelevant just as it does not follow from the fact that some crimes are worse than others that only the worst crimes ought to be illegal.

If these arguments for a general conscience asymmetry are not successful, a few options are available. One option is to argue for a general conscience symmetry. Kyle Fritz takes this route, arguing for conscience symmetry because the same reasons for protecting conscientious objection (e.g., to protect the moral integrity of individual clinicians) also apply to conscientious provision (Fritz 2020). As Fritz notes, an argument for a general conscience symmetry does not specify whether symmetry should be achieved by either 1-rejecting both positive and negative conscience claims or 2-permitting both positive and negative conscience claims.

Another option, and the one defended here, is that only *some* conscientious provisions deserve protection (Brummett et al. 2024). The primary reason for pursuing this more nuanced approach is motivated by an attempt to balance institutional and individual conscience claims based on a critical *practical* difference between objection and provision. Namely, conscientious objections require relatively trivial institutional resources to be honored while conscientious provisions can require significant institutional resources (Brummett 2020b). For example, a physician who refuses to perform a surgical abortion simply recuses themselves from the situation and another physician can then step in to provide the service whereas a physician who wishes to provide a surgical abortion would require time, use of an operating room, staff, equipment, medication, etc. On this analysis, only conscientious provisions that result in trivial institutional burdens at Catholic hospitals that are comparable to the institutional burdens of conscientious objection at secular hospitals should be honored.

One conscientious provision that ought to be protected is the provision of referrals, understood as any verbal or written act that informs the patient where to obtain a medical service or that formally authorizes (for insurance purposes) the patient to see a specialist. Currently, providing a referral to a patient for a medical good or service prohibited by Catholic teaching is prohibited by the ERDs. According to Directive 73, "[A] Catholic institution must ensure that neither its administrators nor its employees will manage, carry out, assist in carrying out, make its facilities available for, *make referrals for*, or benefit from the revenue generated by immoral procedures" (Bishops 2018, 26, emphasis added).

Secular clinicians who work at Catholic hospitals may believe themselves to have a moral obligation to refer patients for procedures prohibited by the ERDs. Secular clinicians may believe that patients have a moral right to receive those treatments, and that a failure to refer makes clinicians morally complicit in violating that right. Such clinicians may also believe that their obligation to refer is increased by their hospital's failure to provide adequate advance notice to patients about prohibited services. For example, a 2019 study of the websites of 646 US Catholic hospitals found that only 28% identified how their religious affiliation could restrict patient access to standard medical care (Takahashi et al. 2019). Providing a referral satisfies the principle of comparatively trivial institutional burdens because it takes very little time to do and requires very little, if anything, in terms of Catholic institutional resources. Therefore, the conscientious provision of referrals ought to be legally protected.

In addition to referrals, the provision of prescriptions for birth control or even physician-assisted-suicide to be filled by a third-party pharmacy may also satisfy the principle of comparatively trivial institutional burdens (Brummett and James 2021). Currently, physicians are prohibited from writing these prescriptions while working in their role at a Catholic institution, and some have been fired for doing so (Ayi 2005). But writing a prescription takes very little time. One study found that physicians need just 11.75 s to enter a prescription through an electronic order-entry system (Coleman et al. 2015). Writing prescriptions also requires little more than use of the electronic medical record. Additionally, Dominic Wilkinson has argued that receiving an ERD-prohibited tubal ligation for an already-scheduled cesarean section requires very low institutional resources and should be a protected form of conscientious provision (Wilkinson 2020). Future research on this topic could be extended to other forms of ERD-prohibited procedures in the attempt to find a balance between institutional and individual conscience following the principle of comparatively trivial institutional burdens.

4.3.2 ERD-Permitted Conscientious Provision

One final form of conscientious provision applies to medical treatments that are not prohibited by the ERDs but may be prohibited by the Catholic hospital where the clinician is employed. For example, some Catholic hospitals prohibit emergency contraception for victims of sexual assault. Recently, 13 Catholic hospitals in Pennsylvania have given notice to the Department of Health that they will not provide emergency contraception (EC) due to a religious or moral belief.[5] In a nationwide study of 597 emergency departments at Catholic hospitals, it was found that 55% would not dispense EC, about half provided a referral, but most referrals did not lead to a willing provider (Harrison 2005). Some Catholic hospitals go so far as to prohibit even discussing EC with rape victims (Smugar et al. 2000).

However, the ERDs do not strictly prohibit prescribing EC in cases of sexual assault. Directive 36 reads, "A female who has been raped should be able to defend herself against a potential conception from the sexual assault. If, *after appropriate testing*, there is no evidence that conception has already occurred, she may be treated with medications that would prevent fertilization. It is not permissible, however, to initiate or to recommend treatments that have as their purpose or direct effect the removal, destruction, or interference with the implantation of a fertilized ovum" (Bishops 2018, 15, emphasis added). Catholic hospitals are free to interpret the "appropriate testing" criterion of Directive 36 as liberally or conservatively as their diocesan bishop will allow. Some bishops may adopt the position of the Catholic Medical Association on EC in cases of rape, which claims that scientific evidence shows the medication may function as an abortifacient, concluding that use of EC is morally illicit (Catholic Medical Association 2016). Among Catholic hospitals that will administer EC, there are conflicting interpretations of what constitutes "appropriate testing" to satisfy Directive 36. Some hospitals use the strict Peoria Protocol, which requires a negative test for ovulation because administering EC prior to ovulation is the only way to rule out the possibility that it could work by preventing implantation of an embryo. A critique of the Peoria Protocol is that it delays the administration of EC and unnecessarily limits Directive 36 because the directive explicitly affirms that medications can be administered to prevent fertilization, which occurs after ovulation. As a result, some Catholic hospitals opt for a less demanding interpretation of "appropriate testing" to require only a simple menstrual history and a pregnancy test, termed the "pregnancy approach" (Hamel and Panicola 2002).

As mentioned above, this essay adopts the reasonability view of conscience such that, to be honored, a conscience claim must be defensible in the public square. Catholic bioethicists have used this standard of reasonable *external* disagreement to defend the conscience rights of Catholic hospitals in secular society (Eberl 2019b). However, this standard can be inverted, such that reasonable *internal* disagreement

---

[5] See: http://www.pacodeandbulletin.gov/Display/pabull?file=/secure/pabulletin/data/vol50/50-6/191.html.

may be used to defend the conscientious provision of EC to rape victims in Catholic hospitals (Brummett et al. 2023). Reasonable internal disagreement occurs when opposing arguments can be given using the resources of Catholic moral teaching for both sides of an issue left open for interpretation by the ERDs. As demonstrated above, the debate over the moral permissibility of prescribing EC to rape victims admits of reasonable internal disagreement over how to interpret "appropriate testing" in Directive 36 of the ERDs. Several concepts internal to the Catholic moral tradition are active in this debate: the moral authority of the Catholic Church, the impermissibility of abortion, moral certainty, the doctrine of double effect, etc. If Catholics are to advocate for protecting conscientious objection in secular society based on reasonable external disagreement, then there is good reason to extend this tolerance to permitting conscientious provision based on reasonable internal disagreement regarding questions left morally open by the ERDs. Because the provision of EC to rape victims is an area of reasonable internal disagreement, Catholic hospitals should allow physician employees to be led by the dictates of their own conscience when choosing which interpretation of "appropriate testing" to follow.

4.4 Conclusion

The conscience debate has typically been framed as an attempt to protect clinicians who wish to refuse legal and professionally accepted medical treatment based on religious beliefs. Some secular bioethicists have responded by arguing for outright rejection of conscience claims in health care, but these arguments have generally failed to produce any legal change. Instead, conscience rights have only grown stronger over time. This essay has taken a different tack by exploring possibilities for secular conscientious objection and provision. If conscience is to be an inevitable feature of health care, then its uses beyond religion deserve sustained exploration.

References

Ashwal, S., R. Cranford, and J. L. Bernat. 1994. The Multi-Society Task Force on PVS. Medical Aspects of the Persistent Vegetative State. *The New England Journal of Medicine* 330:1499–1508.

Ayi, Mema. 2005. *Doctors Fired for Prescribing Birth Control Returninig to Region*. Northwest Indiana Times. https://www.nwitimes.com/news/local/doctors-fired-for-prescribing-birth-control-returning-to-region/article_49984f5a-a773-5576-98b2-2f828e17a8fd.html.

Batavia, Andrew I. 2002. Disability Versus Futility in Rationing Health Care Services: Defining Medical Futility Based on Permanent Unconsciousness—PVS, Coma, and Anencephaly. *Behavioral Sciences and the Law* 20 (3): 219–233.

Bedford, Elliott Louis. 2012. The Concept of Institutional Conscience. *National Catholic Bioethics Quarterly* 12 (3):

Bedford, Elliott Louis. 2016. The Reality of Institutional Conscience. *The National Catholic Bioethics Quarterly* 16 (2): 255–272.

Bishops, United States Conference of Catholic. 2018. *Ethical and Religious Directives for Catholic Health Care Services*. Washington, DC.
Bosslet, Gabriel T., Thaddeus M. Pope, Gordon D. Rubenfeld, Bernard Lo, Robert D. Truog, Cynda H. Rushton, Dee W. Ford, J. Randall Curtis, Molly Osborne, Cheryl Misak, David H. Au, Elie Azoulay, Baruch Brody, Jesse B. Hall, Brenda G. Fahy, Jozef Kesecioglu, Alexander A. Kon, Kathleen O. Lindell, Douglas B. White, and on behalf of The American Thoracic Society ad hoc Committee on Futile and Potentially Inappropriate Treatment. 2015. An Official ATS/AACN/ACCP/ESICM/SCCM Policy Statement: Responding to Requests for Potentially Inappropriate Treatments in Intensive Care Units. *American Journal of Respiratory and Critical Care Medicine* 191 (11): 1318–1330.
Brummett, A. 2020. The quasi-religious nature of clinical ethics consultation. In Hec Forum (Vol. 32, No.3, pp. 199–209). Dordrecht: Springer Netherlands.
Brummett, Abram. 2020a. When Conscientious Objection Runs Amok: A Physician Refusing HIV Preventative to a Bisexual Patient. *Clinical Ethics*: 1477750920946686.
Brummett, Abram L. 2020b. Should Positive Claims of Conscience Receive the Same Protection as Negative Claims of Conscience? Clarifying the Asymmetry Debate. *The Journal of Clinical Ethics* 31 (2): 136–142.
Brummett, Abram, and Jason T. Eberl. 2022. The Many Metaphysical Commitments of Secular Clinical Ethics: Expanding the Argument for a Moral–Metaphysical Proceduralism. *Bioethics* 36 (7): 783–793.
Brummett, Abram, and Eric James. 2021. Non-Catholic Physicians Should be Permitted to Write Prescriptions for Birth Control in Catholic Institutions: Replying to Giubilini and Wilkinson, and Describing a Permissible Positive Conscience Claim. *Journal of Clinical Ethics* 32 (3):
Brummett, Abram, Marlee Mason-Maready, and Victoria Whiting. 2023. Catholic Hospitals Should Permit Physicians to Provide Emergency Contraception to Rape Victims as an Act of Conscientious Provision. *The Linacre Quarterly* 90 (1): 24–34.
Brummett, A., and Eberl, J. 2025. The reasonable content of conscience in public bioethics. *Cambridge Quarterly of Healthcare Ethics 34 (1): 36-48.*
Brummett, A. L., Hafen, T., and Navin, M. C., 2024. Principled Conscientious Provision: Referral Symmetry and Its Implications for Protecting Secular Conscience. *Hastings Center Report 54 (4): 3–10.*
Callahan, Daniel. 1992. When Self-Determination Runs Amok. *The Hastings Center Report* 22 (2): 52–55.
Card, Robert F. 2020. *A New Theory of Conscientious Objection in Medicine: Justification and Reasonability*. Routledge.
Catholic Medical Association. 2016. *Statement on Emergency Contraception in Cases of Rape*.
Chua, Hui Min Hillary. 2020. *Revisiting the Vegetative State: A Disability Rights Law Analysis*.
Coleman, Jamie J., James Hodson, Sarah K. Thomas, Hannah L. Brooks, and Robin E. Ferner. 2015. Temporal and Other Factors That Influence The Time Doctors Take to Prescribe Using an Electronic Prescribing System. *Journal of the American Medical Informatics Association* 22 (1): 206–212.
Crisci, Claudio. 1995. The Ultimate Curse. *Journal of Medical Ethics* 21 (5): 277–280.
Dickens, Bernard. 2014. *The Right to Conscience*. University of Pennsylvania.
Eberl, Jason T. 2019a. Conscientious Objection in Health Care. *Theoretical Medicine and Bioethics* 40 (6): 483–486.
Eberl, Jason T. 2019b. Protecting Reasonable Conscientious Refusals in Health Care. *Theoretical Medicine and Bioethics*1–17.
Edlow, Brian L., Jan Claassen, Nicholas D. Schiff, and David M. Greer. 2021. Recovery From Disorders of Consciousness: Mechanisms, Prognosis and Emerging Therapies. *Nature Reviews Neurology* 17 (3): 135–156.
Fox, Dov. 2022. *Medical Disobedience. San Diego Legal Studies Paper (22-028).*
Fritz, Kyle G. 2020. Unjustified Asymmetry: Positive Claims of Conscience and Heartbeat Bills. *The American Journal of Bioethics* 1–20.

Giacino, Joseph T., Douglas I. Katz, Nicholas D. Schiff, John Whyte, Eric J. Ashman, Stephen Ashwal, Richard Barbano, Flora M. Hammond, Steven Laureys, and Geoffrey S. F. Ling. 2018. Practice Guideline Update Recommendations Summary: Disorders of consciousness: Report of the Guideline Development, Dissemination, and Implementation Subcommittee of the American Academy of Neurology; the American Congress of Rehabilitation Medicine; and the National Institute on Disability, Independent Living, and Rehabilitation Research. *Archives of Physical Medicine and Rehabilitation* 99 (9): 1699–1709.

Hamel, Ronald P., and Michael R. Panicola. 2002. Emergency Contraception and Sexual Assault. *Health Progress* 83 (5): 12–13.

Harrison, Teresa. 2005. Availability of Emergency Contraception: A Survey of Hospital Emergency Department Staff. *Annals of Emergency Medicine* 46 (2): 105–110.

Hershenov, David. 2021. Conscientious Objection or An Internal Morality of Medicine? *Christian bioethics: Non-Ecumenical Studies in Medical Morality* 27 (1): 104–121.

Johnson, L. Syd M. 2022. *The Ethics of Uncertainty: Entangled Ethical and Epistemic Risks in Disorders of Consciousness*. Oxford University Press.

Kon, Alexander A., Eric K. Shepard, Nneka O. Sederstrom, Sandra M. Swoboda, Mary Faith Marshall, Barbara Birriel, and Fred Rincon. 2016. Defining Futile and Potentially Inappropriate Interventions: A Policy Statement From the Society of Critical Care Medicine Ethics Committee. *Critical Care Medicine* 44 (9): 1769–1774.

Ludka, N., Hurse, D., and Brummett, A. 2025. An investigation into the public's attitude toward opting out of brain death. *Neurocritical Care 43 (1): 262–276.*

Miller, Franklin G., and Howard Brody. 1995. Professional Integrity and Physician-Assisted Death. *Hastings Center Report* 25 (3): 8–17.

Montagnino, Barbara A., and Angela M. Ethier. 2007. The Experiences of Pediatric Nurses Caring for Children in a Persistent Vegetative State. *Pediatric Critical Care Medicine* 8 (5): 440–446.

Monti, Martin M., Audrey Vanhaudenhuyse, Martin R. Coleman, Melanie Boly, John D. Pickard, Luaba Tshibanda, Adrian M. Owen, and Steven Laureys. 2010. Willful Modulation of Brain Activity in Disorders of Consciousness. *New England Journal of Medicine* 362 (7): 579–589.

Payne, Kirk, Robert M. Taylor, Carol Stocking, and Greg A. Sachs. 1996. Physicians' Attitudes About the Care of Patients in the Persistent Vegetative State: A National Survey. *Annals of Internal Medicine* 125 (2): 104–110.

Pope, Thaddeus. 2023. *Exemptions from Brain Death: Why the Uniform Law Commission Recommends It*. Medical Futility Blog. https://medicalfutility.blogspot.com/2023/06/exemptions-from-brain-death-why-uniform.html.

Rawls, John. 1972. *A Theory of Justice*. Oxford: Clarendon.

Rhodes, Rosamond. 2019. Conscience, Conscientious Objections, and Medicine. *Theoretical Medicine and Bioethics* 40 (6): 487–506.

Rich, Ben A. 1997. Prospective Autonomy and Critical Interests: A Narrative Defense of the Moral Authority of Advance Directives. *Cambridge Quarterly of Healthcare Ethics* 6 (2): 138–147.

Scherer, Liz. 2022. *When HIV Stigma and Conscientious Objection Collide*. Medscape. https://www.medscape.com/viewarticle/978615#vp_2.

Schiff, Nicholas D., Joseph T. Giacino, and Joseph J. Fins. 2009. Deep Brain Stimulation, Neuroethics, and the Minimally Conscious State: Moving Beyond Proof of Principle. *Archives of Neurology* 66 (6): 697–702.

Schneiderman, Lawrence J., and Nancy S. Jecker. 2011. *Wrong Medicine: Doctors, Patients, and Futile Treatment*. JHU Press.

Schuklenk, Udo. 2019. Conscience-Based Refusal of Patient Care in Medicine: A Consequentialist Analysis. *Theoretical Medicine and Bioethics*1–16.

Sepper, Elizabeth. 2012. Taking Conscience Seriously. *Virginia Law Review* 98:1501.

Shamus, K. J. 2020. *Titus Cromer, Teen at Center of Life Support Legal Battle, Dies at Age 16*. Detroit Free Press.

Smugar, Steven S., Bernadette J. Spina, and Jon F. Merz. 2000. Informed Consent for Emergency Contraception: Variability in Hospital Care of Rape Victims. *American Journal of Public Health* 90 (9): 1372.

Takahashi, Joelle, Abigail Cher, Jeanelle Sheeder, Stephanie Teal, and Maryam Guiahi. 2019. Disclosure of Religious Identity and Health Care Practices on Catholic Hospital Websites. *JAMA* 321 (11): 1103–1104.

Truog, Robert D. 2018. Defining Death—Making Sense of the Case of Jahi McMath. *JAMA* 319 (18): 1859–1860.

Veatch, Robert M., and Lainie Friedman Ross. 2016. *Defining Death: The Case for Choice*. Georgetown University Press.

Wasserman, J. A., Brummett, A., and Navin, M. C. 2026. A defense of institutional conscience rights for secular hospitals: philosophical justifications and practical applications. *The American Journal of Bioethics 26 (7): 45–58.*

Wicclair, Mark R. 2009. Negative and Positive Claims of Conscience. *Cambridge Quarterly of Healthcare Ethics* 18 (1): 14–22.

Wicclair, Mark R. 2011. *Conscientious Objection in Health Care: An Ethical Analysis*. Cambridge University Press.

Wildes, Kevin Wm. 1997. Institutional Identity, Integrity, and Conscience. *Kennedy Institute of Ethics Journal* 7 (4): 413–419.

Wilkinson, Dominic Jc. 2020. Positive or Negative? Consistency and Inconsistency in Claims of Conscience. *The Journal of Clinical Ethics* 31 (2): 143–145.

Zandbergen, E. G. J., A. Hijdra, J. H. T. M. Koelman, A. A. M. Hart, P. E. Vos, M. M. Verbeek, and R. J. De Haan. 2006. Prediction of Poor Outcome Within the First 3 Days of Postanoxic Coma. *Neurology* 66 (1): 62–68.

Chapter 5
Conscientious Objection in Healthcare

Xavier Symons

Abstract Freedom of conscience is a fundamental ethical and political principle in modern liberal democracies. The practice of conscientious objection, however, has been a source of extensive debate in recent years, particularly in the sphere of healthcare. This article offers a defense of conscientious objection grounded in both subjective and objective features of moral conscience. I argue that conscientious objection is (1) a public act that (2) is plausibly based on a virtuous search for moral truth, and that (3) represents a deeply held belief of the moral agent. Given this definition, there ought to be a presumption in favor of permitting conscientious objection within the health professions.

5.1 Introduction

Freedom of conscience is a political and legal principle at the heart of the modern liberal democratic state. Respect for individual conscience is precisely one hallmark of what makes a democracy *liberal* and not a mere tyranny of the democratic majority or prevailing public opinion. Conscientious objection, however, in its various manifestations in civil society and professional life, has been subject to extensive criticism and remains contentious today. Recent scholarship on conscientious objection has focused on the permissibility of conscientious objection in healthcare. Some scholars have argued for extensive restrictions on conscientious objection in medicine and other health professions (McLeod 2008; Card 2020), and some have gone as far as to argue for an outright ban (Savulescu 2006). This essay, however, offers an ethical defense—grounded in an articulation of both objective and subjective aspects of moral conscience—of conscientious objection in healthcare. I argue that conscientious objection is (1) a public act that (2) is plausibly based on a virtuous search for moral truth, and that (3) represents a deeply held belief of the moral

X. Symons (✉)
Harvard Medical School, Center for Bioethics, Boston, MA, USA
e-mail: xavier_symons@hsph.harvard.edu

K. R. Smith (ed.), *Bioethics and Religion*, The International Library of Bioethics 116, https://doi.org/10.1007/978-3-032-33538-8_5

agent. Given this definition, conscientious objection is worthy of presumptive respect and tolerance within the health professions. This argument is principally focused on individual healthcare practitioners but an argument by extension can be made to the right of healthcare institutions to conscientiously object (Symons and Chua 2024).

This chapter will begin with some general reflections on conscience and proceed to a more particular discussion of conscientious objection. The first section of this chapter defines conscience, while section two offers a definition of conscientious objection. Part three offers an overview of conscientious objection, while part four offers a critical evaluation of arguments for and against conscientious objection and part five briefly explores the notion of institutional conscientious objection. This essay closes with some brief policy recommendations for the regulation of conscientious objection in healthcare.

5.2 What Is Conscience?

This section offers an overview of conscience as a moral psychological entity. Conscience has several dimensions, though fundamentally it is a native capacity of the human mind that allows one to determine right and wrong and to assess how the obligations of morality apply to particular situations. Conscience, in its most general sense, refers to the moral capacity to distinguish right and wrong, good and evil, and praiseworthy and blameworthy conduct. It is a capacity for *judgment* insofar as its primary function is to pass moral judgment on actions, notwithstanding any secondary function it may play in helping one maintain a moral identity and stay true to one's deep commitments.

Plausibly, conscience has an affective dimension insofar as it is linked to moral emotions like guilt or shame. But crucially, these features of moral conscience arise from a prior judgment about the rightness or wrongness of actions and conscience ought not be reduced to mere sentiment. Thus, *guilt* is posterior to a judgment that one has acted wrongly against another person and that one ought to make reparation for wrongdoing. *Shame* is a moral emotion that captures the sensation of being perceived to have acted wrongly by others. We might call shame an imaginative feature of conscience insofar as it involves an imaginative awareness of real or perceived observers of our moral conduct.

While conscience is viewed by some as irredeemably subjective, early pagan and Christian accounts of conscience sought to preserve a sense of the objectivity of the judgements of conscience. Thus, conscience has been seen in both Classical pagan and Christian contexts as *the moral law within us*. Stoics such as Epictetus believed that we have certain natural pre-conceptions of good and evil, while Cicero held that we have innate within us a grasp of specific moral rights, such as a right to self-defense (Sorabji 2014). St Paul writes of a law written on the hearts of Gentiles that guides them even when they are not possessed of revealed truth (*Rom* 2:14–15). Importantly, for St Paul conscience is a "co-witness" to the law written on our hearts and, as such, Pauline conscience would appear to be a judgment of practical reason based on the objective moral law given innately to us by God.

For Thomas Aquinas, conscience is a judgment of practical reason whereby we apply general moral principles to particular cases. But there is also a sense in which conscience can refer to *synderesis*, or the first principle of practical reason, namely, that good ought to be pursued and that evil ought to be avoided. While considered in itself *synderesis* as the most general of moral principles may seem to lack specific moral content, according to some commentators, at least, it includes an inchoate knowledge of the good (Knobel 2021). As such, conscience *qua synderesis* is anchored in a transcendent and objective good, even if the antecedent principles that one draws from it may depart from the objective good (Finnis 1979).

More recent secular accounts of conscience tend to focus on its importance in moral development and the maintenance of our subjective identity commitments. Freud equated conscience with the super-ego and viewed conscience as, among other things, an interiorisation of the moral censures of one's parents. This, rather than the consistent rational functioning of the mind, was the source of moral principles like the categorical imperative (Velleman 1999). Given its association with parental love and sanction, *"the voice of conscience is an affective one, comprising the emotions of anxiety, shame, guilt and remorse"* (p. 485) (Lyons 2009).

This should suffice for an overview of conscience as a moral psychological concept. Some commentators may, however, question the necessity of a robust account of conscience for discussions of conscientious objection in healthcare, which is the focus of this essay. One might be skeptical about the very existence or reality of conscience. One might wonder whether we ought instead to rely on a less controversial moral psychological vocabulary including terms like moral reason or moral convictions. Thus, some might focus on the deeply held beliefs of moral actors or perhaps even the moral significance of identity or integrity when explaining why it is problematic for a moral agent to act against their conscientious commitments.

The trouble with 'conscience skepticism'—as we might call this view—is that it overlooks the rich interrelations that exist between moral reason, moral emotions, and moral identity. Perhaps the strongest argument for the existence of conscience is its explanatory power for understanding how acting against a deep belief can trigger moral emotions like guilt and shame and also feelings of self-betrayal. It is difficult to capture the phenomenology of moral experience, particularly guilt, shame, and a sense of moral disintegration, without recourse to the language and dynamics of conscience.

5.3 Why Ought We Respect Conscience?

Why, from a philosophical perspective, ought we respect conscience? In this section, I will advance two complementary reasons for respecting conscience. First, conscience tracks objective morality. Second, action in conformity with conscience demonstrates soundness of the will. These two reasons for respecting conscience are closely related but also distinct in important ways.

Respect for conscience as a political principle is enshrined in several national and international codes of rights. The International Charter on Civil and Political Rights—ratified by nations such as Australia, Canada and the United States—recognises a 'right to freedom of thought, conscience and religion' (United Nations (General Assembly) 1966). This right is also enshrined in the European Union Charter of Fundamental Rights (European Union 2010). But the basis for this right is widely debated. Hence in this section, I will offer a philosophical justification for respecting individual consciences.

The first and primary reason for respecting conscience is that conscience tracks objective morality. A well-functioning conscience is what the scholastics called "right reason in action" or *prudentia* and ought to be respected insofar as it is reflective of a sound application of universal moral principles to particular situations. This idea presupposes—not uncontroversially—that conscience is actually tracking objective morality. It necessitates that moral agents are making an effort to get at the truth of morality and form their consciences appropriately. Reason, after all, is a capacity whereby we approximate reality with our minds, and thus truth is defined in scholastic thought as *adequatio rei et intellectus*—the agreement between the thing and the intellect (*Summa Theologiae* Question XVI, Art. 1, 3) (Saint Thomas 1920).

Unsurprisingly, many push back against the idea that conscience tracks moral truth. For one, conscience would appear to lead one to not only manifestly erroneous conclusions but also downright prejudicial moral positions. Consider, by way of example, Mark Twain's novel *Huckleberry Finn*, where Huck, a 13-year-old white teenager, is travelling on a raft with Jim, an African American adult slave who has run away from his owner. Huck is emotionally torn about whether to turn Jim in or whether to assist him in his escape. Huck feels like he has a duty to turn Jim in, and indeed this sense of duty is described by Twain as originating from the boy's conscience. Huck feels like it would be a sin to not return Jim to his 'rightful' owner. But Huck also has an intense, intuitive feeling of unease about betraying Jim and ends up refraining from doing so (Chap. 37) (Twain 1994). Conscience would not seem to be tracking objective morality in this case, and it seems difficult to see why conscience should be respected on the basis of some purported connection with objective morality. On the contrary, if anything it seems that conscience would need to be retooled and disabused of its racial and other prejudices (Brownlee 2012; McLeod 2020).

But this objection confuses conscience's orientation to objective morality with the misguided notion of conscience being infallible. On the contrary, conscience is oriented toward moral truth but can malfunction and in some cases can be radically deformed by community and social influences. The possibility of conscience being deformed is evident from experience and is also acknowledged in a Catholic Christian view of conscience. Thus, the *Catechism of the Catholic Church* (Catechism of the Catholic Church 1997) speaks of the grave importance of the proper formation of conscience over the course of one's lifetime:

> The education of conscience is indispensable for human beings who are subjected to negative influences and tempted by sin to prefer their own judgment and to reject authoritative teachings...The education of the conscience is a lifelong task...in the formation of conscience [...] we are assisted by the gifts of the Holy Spirit, aided by the witness or advice of others and guided by the authoritative teaching of the Church. (n. 1783–1785)

It does not follow, then, from the fact that conscience can be deformed that conscience possesses no orientation to moral truth. On the contrary, the possibility of an erroneous conscience provides further reason to take seriously the duty to form one's own conscience. Respect for conscience ought always be complemented by an emphasis on the formation of conscience as a grave personal and societal responsibility.

Notwithstanding the possibility of a deformed conscience, then, conscience ought to command a certain social and political respect given its orientation toward moral truth. Thus, John Paul II wrote in the Encyclical *Veritatis Splendor* that *"it is always from the truth that the dignity of conscience derives"* (n. 62) (Catholic Church 1993). When functioning correctly, conscience gives voice to the moral law and so is deserving of obedience by a moral agent and respect from the society in which that agent lives and acts.

This position helps us respond to the view that conscience leads us into a hopeless form of relativism. Savulescu and Schuklenk, for example, have cautioned that "[a]t the heart of the paradox of conscientious objection is arguably a belief in ethical relativism" (p. 167) (Savulescu and Schuklenk 2017). The account of respect for conscience that I have offered, however, grounds respect for conscience in the orientation of conscience toward objective categories of right and wrong and good and evil. As such, it avoids criticisms of ethical relativism.

That being said, respect for conscience need not be limited to situations where conscience judges correctly. On the contrary, conscience ought to be followed when it is in error and still has a claim on social and political respect even in these circumstances. To put the view in classical terms, *an erring conscience binds* notwithstanding the error in one's own moral reasoning. Aquinas was well known for defending this view. Aquinas's view was that a moral agent does something that is subjectively wrong in acting contrary to conscience even if their resulting action is objectively right. There is a goodness in the will even if reason is perverse. Summarizing Aquinas's view, Philippa Foot (2001) writes:

> ...even an erring conscience binds, [because] in going against conscience the will tends to an action as something evil in that reason has proposed it as evil. And surely this is right. For acting as one thinks one should not is a very radical form of badness in the will. How could a human being be acting well in doing what he or she saw as evil (pp. 73–74).

A consequence of Aquinas' view is that a person with an erring conscience can find themselves unwittingly in a damnable moral situation, where if they do the objectively right thing they betray conscience, or if they follow conscience they do a morally wrong act. Crucial for our purposes, however, is the fact that the moral agent does not see the damnable situation in this way. The agent only has their conscience to go by and, as such, believes in good faith that they act rightly.

A critic might argue that, regardless of the soundness of one's will, morally abhorrent acts are not deserving of our respect. Thus, a Neo-Nazi who in accord with their distorted conscience seeks to persecute people of Jewish and black ethnicity ought to be apprehended and subject to the full force of the law. For argument's sake we can set aside situations where an agent is not acting in accord with any well-worked out, rationally consistent account of morality and is rather totally wanton toward morality. Instead, we might think of an example where someone has adopted a comprehensive racist ideology and is genuinely convicted of this prejudicial worldview. The central protagonist of the 1998 film *American History X*, Derek Vinyard, might be an example of this. Derek Vinyard begins his adult life as a fervent neo-Nazi and white supremacist, deeply entrenched in the ideology of hate and racial superiority. Derek's worldview is shaped by the murder of his father, which he believes was committed by black drug dealers. This event fuels his anger and leads him down a path of bigotry and violence. Derek is charismatic and articulate, able to articulate his beliefs with conviction, which attracts followers to his cause. He is portrayed as highly intelligent but also deeply troubled, struggling with inner demons and a sense of injustice.

One might argue that a case like this illustrates how someone can have a very intelligent and well-worked out moral view of the world that is nevertheless manifestly abhorrent and undeserving of respect or subtle assessment on our part. As such, we need not respect those who have an erring conscience and believe that their perverse actions are moral.

But note that in Vinyard's case, Aquinas's point remains that "acting as one thinks one should not is a very radical form of badness in the will". Our focus, to be clear, is on the will, not on the rational justification that someone like Vinyard provides for their actions. Thus it would be worse for him to abrogate his moral reason altogether; perhaps if he had done this, and simply been a blind conformist to the ideology that had been foist upon him in adolescence, he may not have undergone the radical transformation that he does later in the narrative (I will not say more at risk of spoiling the story for those who have not seen the film). Suffice to say that even in a situation of an erring conscience we ought to recognise the virtue of the will of the agent to act in accord with right reason even if one's moral judgements are horribly mistaken.

Another question is whether, all things considered, an erring conscience ought to receive political and legal respect. In the case of Derek Vinyard, it seems like the answer is no, at least insofar as his conscience leads him to engage in hate speech or violent behavior against Jews or people of color. But there may be other situations where an individual whose conscience errs ought to be allowed to act on their conscience. We will discuss the ethics of conscientious objection in the later sections of this chapter but suffice to say that relevant factors to consider would be the impact of one's actions on others and on the moral atmosphere of society.

This should suffice for a discussion of soundness of will as a basis for respect for conscience. The overarching purpose of this excursus is to show that respect for conscience has a dual basis in conscience's link to moral truth and its claim to obedience on the good will of moral agents. The free exercise of conscience is important both because it is a tracker of the truth and because conscientious behavior is indicative of a good will.

Importantly, the discourse on conscience rights has evolved over time to emphasize the latter reason over the former. It is the notion of judgements of conscience as subjectively binding that today is most commonly used as a justification for respect for conscience. Recent legal jurisprudence gives witness to the turn from conscience as a source of moral insight to conscience as revelatory of deep, personal, identity commitments. Stephen Smith glosses 100 years of recent jurisprudence with the claim that: *"the protections of conscience in the modern view reflects a recognition of the inviolability or sanctity of individual human beings. Or, we might say, the sanctity of the self"* (Smith 2023). There are problems with this, however, as it seems that conscience's orientation toward objective truth is the source of our reason to respect it. In the words of Newman, *"conscience has rights because it has duties"*, and the ultimate source of the binding character of those duties is found in a moral law that transcends the self and toward which we are called to conform ourselves (Newman 1875).

5.4 What Is Conscientious Objection?

Conscientious objection historically has been associated with religious opposition to public health mandates and, perhaps most significantly, refusal of individuals to participate in what are deemed to be unjust wars. As I have noted elsewhere, historically, the term "conscientious objector" was first applied to a group of citizens in late nineteenth-century England who refused vaccination for smallpox. The British Vaccination Act of 1898 provided a conscience clause to allow exemptions to mandatory smallpox vaccination, thus giving rise to the term "conscientious objector" that we use today (Wolfe and Sharp 2002). In the twentieth century conscientious objection also came to be associated with a refusal to fight in war. In the first, second and Vietnam Wars, conscientious objection was a contentious topic that was in many cases dealt with legal penalties by governments. Conscientious objectors, however, in some cases received significant public support (this was especially true of the Vietnam War).

In more contemporary contexts, conscientious objection has been a widely debated topic in the health professions. In the health professions, is it not uncommon that a health professional will opt out of participation in procedures and practices to which they have a moral objection. 'Conscience clauses' are present in legislation at both a state and federal level in the United States and conscientious objection is also circumscribed in bioethics legislation on abortion and euthanasia in jurisdictions around the world. Medical Associations at a global and national level also provide qualified recognition and defense of the institution of conscientious objection.

It is important to distinguish conscientious objection from civil disobedience. In the most general sense, conscientious objection refers to a refusal to participate in an action that one deems to be immoral. We might usefully distinguish *conscientious objection*, which has as its focus right action and the preservation of moral integrity, from *civil disobedience*, which has as its focus the undermining of a law or regime that one deems to be unjust. While conscientious objection is a public act, insofar as one opts-out of participation in some professional or political practice, it can be distinguished from civil disobedience insofar as civil disobedience contains within it a desire to undermine the law and a desire to be seen experiencing the consequences of undermining the law. Civil disobedience does not seek an exemption from the law; it seeks to undercut the law as such. Thus, Joseph Raz (1979) defines civil disobedience as:

> [the] politically motivated breach of law designed either to contribute directly to a change of a law or of a public policy or to express one's protest against, and dissociation from, a law or a public policy (p. 263).

Conscientious objection, in contrast, need not be aimed at undermining the law. It may be that conscientious objection is explicitly accommodated by the law, as is the case with bioethical law in many jurisdictions governing, for example, abortion and euthanasia. Nor need conscientious objection be politically motivated. As we shall see, conscientious objection can be motivated by a variety of religious and ethical considerations independent of any political considerations.

Kimberley Brownlee argues that civil disobedience may in fact be more defensible insofar as the 'privacy' of conscientious objection may cast doubt on the sincerity of a conscientious objector's commitment:

> [U]nlike private "conscientious" objectors, civil disobedients are willing to risk being seen, and thus held to account, for their conscientious disobedience. As such, their acts do not raise the specter of doubt that conscientious objectors' acts can raise that their conviction is too shaky to accept the risks of communication (2012, p. 19).

I would push back against Brownlee, however, and argue that in fact conscientious objectors are often acting in a public way and in full public view even if they do not intend to undermine the functioning of the law. As such, conscientious objectors ought not be viewed as inauthentic or unwilling to face intense public scrutiny if necessary.

5.5 A Defense of Conscientious Objection

And so we arrive at the main claim of this essay. This is, namely, that if conscientious objection is (1) a public act that (2) is plausibly based on a virtuous search for moral truth, and that (3) represents a deeply held belief of the moral agent, it is deserving of presumptive respect. There may of course be cases where this presumption may be defeated. But, to put it succinctly, it is a case of presumed innocence until proven guilty, and not the reverse.

Some may think that conscientious objection is damaging to social order and cohesion and, as such, ought to be restricted if not outlawed entirely. At the very least, the default ought to be that one obey the law and only in exceptional circumstances ought one be allowed to opt out or be relieved of one's societal or political responsibilities. Consider, for example, military service, which is still practiced in jurisdictions such as Singapore and Taiwan. It could be argued that conscientious objection must be restricted in these contexts or the very institution of mandatory military service could be undermined, as too many people would take advantage of the right to opt-out and it would be unfair to those who *do* participate in military service.

I have argued, however, that we have both subjective and objective reasons for respecting the right to conscientious objection. The robust functioning of conscience implicit in conscientious objection reflects a moral agent's search for the truth and also gives voice to one's deepest commitments. We ought to recognise appeals to conscience for what they are—a reflection of the proper functioning of practical reason and an effort to preserve moral integrity.

It is also worth noting that conscientious objection as a tolerated social practice reflects an acknowledgement of common and profound social disagreement about the collective actions of a polity and the desire to avoid coercing people who have a moral or religious objection to participation in these collective actions. Conscientious objection's *raison d'etre* as an institution is the accommodation of ethical and religious disagreement among society. Thus, in an article published during the time of the Vietnam War, the philosopher Carl Cohen (2014) wrote:

> Provision for conscientious objection is a mark of considerable sophistication in a political community. It indicates that there is a general awareness, in that community, of the depth of moral disagreements, and it is a concrete exhibition of the community's disposition to tolerate such disagreements when they are genuine and profound.

This is a liberal democratic defense of conscientious objection. Cohen's claim is that conscientious objection exists for the sake of acknowledging and to some extent resolving the presence of profound moral disagreements within a polity in a manner that does not cast one side or other as fundamentally unreasonable or that fails to give political or legal recognition to their considered views.

But let us move to a more explicit focus on medicine. As per the difference between conscientious objection and civil disobedience, we can note that the typical expectation is that conscientious objectors will not undermine the provision of the service in question and, in some cases, that the conscientious objector will advise the patient on how they might access the service (though this requirement is controversial). This distinguishes it from an attempt to undermine the law insofar as the conscientious objector is not seeking to undercut the provision of the relevant medical service.

Some scholars think that conscience provisions as such are problematic. Various objections are put forward in the literature but fundamentally the claim is that conscientious objection in medicine conflicts with the professional responsibilities of doctors. Doctors ought to be fully committed to care for their patients and conscientious objection can interfere with this commitment. At an extreme, we might see conscientious objection as a form of abandonment of patients by the one who ought to be their fiduciary. Or we might see it more generally as undermining the promotion of the welfare and rights of patients.

But this argument risks begging the question. Conscientious objection is an institution that reflects an acknowledgement of moral disagreement within society about precisely social practices the morality of which is strongly contested. We ignore this when we presume, say, that abortion is healthcare or medical aid in dying is part of a basic suite of ordinary end-of-life care options.

In keeping with its distinction from civil disobedience, conscientious objection ought to be seen as a shield rather than a sword. That is to say, conscientious objectors ought not see conscience provisions as an opportunity to frustrate the provision of controversial medical services in a particular health facility. This would lapse into civil disobedience, which as a practice would seem to have a higher bar of justification. If indeed conscientious objectors confine their actions to this defensive posture, it is hard to see how conscientious objection would threaten patient rights, though there is some literature that seeks to evidence such a claim.

An alternative concern about conscientious objection is that it can undermine equitable service provision of medical services. In the case of abortion and euthanasia some clinicians fear that conscientious objection in rural areas can severely limit access to these services in those areas. But patients in theory ought to be able to access care elsewhere. Doctors should not abandon their patients if they have a conscientious objection. They should ensure that their patients receive all other forms of care provision, and, if necessary, the patient be provided with information about other care teams should the conscientious objection prove to be a fundamental obstacle to ongoing care.

In addition, there is a need for transparency when one conscientiously objects. That is to say, one could indicate to health authorities that one is a conscientious objector in places where health access is limited and health authorities could seek to ensure adequate medical support is available for patients in that area notwithstanding an individual's conscientious objection. While individual practitioners should not be indifferent to macro-level questions of healthcare access, the primary body responsible for equitable health care access is the State.

Some critics of conscientious objection make an epistemic argument about who is the proper interpreter of the nature of health and the role of health professionals. It could be argued that the profession in general, rather than particular health professionals, is best positioned to determine the scope and limits of the primary interest of doctors.

Once again, this argument seems to ignore the *raison d'etre* for conscientious objection. Thus scholars have argued that conscientious objection exists in part as a recognition of the fallibility of professional associations and the law. Sometimes professional associations and the law can get things wrong. This is evidenced by the history of racism in medicine as well as the widely accepted practice of eugenics in the United States in the early part of the twentieth century. Conscience might be conceived as a check on error in the profession, and a counterbalance to professional consensus on controversial moral matters.

This points us back to the deeper dimension of the argument in this chapter—that conscience tracks objective morality. The horizons of conscience go beyond professional standards. Note that this is a more ultimate standard of morality and deserves a kind of deference, notwithstanding the importance of professional norms. Indeed, medicine ought to recognise a tension between its internal moral code and the sources of normativity on which agents draw outside of medicine. Individual practitioners ought to make some accommodation for patients in leaving aside their own personal views in the clinic. But cases of conscientious objection are a limiting case on this principle. Individual practitioners should be able to opt out of care when their conscientious commitments are at stake.

In summary, conscientious objection is presumptively justified, both specifically in medicine and more generally in other societal contexts. This justification is based on the fact that conscientious objection is (1) a public act that (2) is plausibly based on a virtuous search for moral truth, and that (3) represents a deeply held belief of the moral agent.

We might close by considering that conscientious objection is not unrelated to the cultivation of character and virtue in medicine. Physicians need to learn how to exercise their moral judgment, and perhaps one of the places in medicine where this judgment is most keenly refined is in the most controversial aspects of medical practice, which is precisely the matter of conscientious objection. Paradoxically, we need to give doctors the possibility of getting things wrong if we want them to be able to independently judge what is right. The mere permissibility of conscientious objection goes some way toward cultivating this capacity for independent clinical and moral judgment.

5.6 Institutional Conscientious Objection

Lastly, we will consider whether there is an analogy between institutional and individual conscience. It is our contention that there is good reason to accommodate institutional conscientious objection. This section will necessarily be brief as a comprehensive discussion of this topic would require a chapter in itself.

Granted, institutions may not have consciences in the literal way that we attribute consciences to individuals. Institutions are not themselves psychologically aware. There are some scholars who claim this. But institutions do not have minds. Rather, we can see institutions as groups of people acting in concert to realize goods that are not realizable by any one individual. We can also attribute a kind of collective exercise of conscience to institutions. In this way, we can say that institutions possess what we might call *conscience**. There is something analogous to the pursuit of institutions of a particular mission grounded in the common good and the pursuit of individuals of moral truth.

Insofar as institutions can be said to be analogous to individuals, they are deserving of *conscience rights**. We say this for several reasons: respect for institutional conscience is good for the institutions themselves; it's good for doctors; it's good for patients; and it's good for society generally. It's good for institutions because it helps preserve their integrity. It's good for doctors and patients because it provides havens where doctors and patients can go where they know there will not be certain procedures performed. It's good for society because society needs institutions that are strong and that can provide the services that only institutions and not individuals alone can provide. This is not an exhaustive list of the reasons to respect institutional conscience rights but it's a good starting point.

Critics, of course, will caution of the damage in some cases that institutions have brought about in societies. But we ought not let exceptions dictate the rules. Institutions are in the vast majority of cases common goods—i.e., goods realised through collective action and that are conducive to the welfare and flourishing of all involved—and we ought to seek to support and preserve these common goods. Institutions that conscientiously object, then, are worthy of support and preservation if only because the institutions themselves are an important part of a flourishing society and the harmonious cooperation of the State with civil society.

5.7 Conclusion

We can close by considering the policy implications of the argument presented in this chapter. The tendency in bioethics literature has been toward placing significant restrictions on conscientious objection. Some have gone as far as to argue for a ban on conscientious objection. Indeed, some laws have restricted or prohibited certain forms of conscientious objection. I would argue that both individual and institutional conscientious objection ought, as a rule, to be respected. Individual conscientious objection is worthy of respect as insofar as conscience represents someone's best efforts to get at moral truth. Institutional conscience is worthy of respect insofar as institutions are important social goods and require some degree of autonomy to practice in accord with their mission and values. This principle has limits, but it applies at least as a rule where some of the most contentious practices in medicine, both at the beginning and end of life, are concerned.

References

Brownlee, Kimberley. 2012. *Conscience and Conviction: The Case for Civil Disobedience*. Oxford University Press UK.

Card, Robert F. 2020. *A New Theory of Conscientious Objection in Medicine: Justification and Reasonability*. Routledge.

Catechism of the Catholic Church. 1997. 2nd ed. Vatican City: Libreria Editrice Vaticana. http://www.vatican.va/archive/ENG0015/_INDEX.HTM.

Catholic Church. 1993. *Veritatis Splendor: Encyclical Letter Addressed by the Supreme Pontiff Pope John Paul II to All Bishops of the Catholic Church Regarding Certain Fundamental*

Questions of the Church's Moral Teaching. London: Incorporated Catholic Truth Society in collaboration with Veritas Publications, Dublin.
Cohen, S. Robin. 2014. McGill Quality of Life Questionnaire. In *Encyclopedia of Quality of Life and Well-Being Research*, edited by Alex C. Michalos, 3883–3889. Dordrecht: Springer Netherlands. https://doi.org/10.1007/978-94-007-0753-5_1749.
European Union. 2010. *Charter of Fundamental Rights of the European Union. Official Journal of the European Union C83*. Vol. 53. Brussels: European Union.
Finnis, John. 1979. *Natural Law and Natural Rights*. Vol. 31. Oxford University Press UK.
Foot, Philippa. 2001. *Natural Goodness*. Oxford University Press.
Knobel, Angela Mc Kay. 2021. *Aquinas and the Infused Moral Virtues*. Notre Dame Press.
Lyons, William. 2009. Conscience – An Essay in Moral Psychology. *Philosophy* 84 (4): 477–494.
McLeod, Carolyn. 2008. Referral in the Wake of Conscientious Objection to Abortion. *Hypatia* 23 (4): 30–47.
McLeod, Carolyn. 2020. *Conscience in Reproductive Health Care: Prioritizing Patient Interests*. Oxford University Press.
Newman, John Henry. 1875. *A Letter Addressed to His Grace the Duke of Norfolk: On Occasion of Mr. Gladstone's Recent Expostulation*. Sydney: F Cunninghame & Co.
Raz, Joseph. 1979. *The Authority of Law: Essays on Law and Morality*. Vol. 3. Oxford University Press.
Saint Thomas, Aquinas. 1920. *The "Summa Theologica" of St. Thomas Aquinas ...* London: Burns, Oates & Washburne, Ltd., [1920-42]. https://search.library.wisc.edu/catalog/999609076002121.
Savulescu, Julian. 2006. Conscientious Objection in Medicine. *BMJ (Clinical Research Ed.)* 332 (7536): 294–297. https://doi.org/10.1136/bmj.332.7536.294.
Savulescu, Julian, and Udo Schuklenk. 2017. Doctors Have No Right to Refuse Medical Assistance in Dying, Abortion or Contraception. *Bioethics* 31 (3): 162–170. https://doi.org/10.1111/bioe.12288.
Smith, Stephen. 2023. *The Distintegrating Conscience and the Decline of Modernity*. South Bend, IN: University of Notre Dame Press Notre Dame, Ind.
Sorabji, Richard. 2014. *Moral Conscience Through the Ages: Fifth Century Bce to the Present*. University of Chicago Press.
Symons, Xavier, and Reginald Mary Chua. 2024. Three Arguments Against Institutional Conscientious Objection, and Why They Are (Metaphysically) Unconvincing. *Journal of Medicine and Philosophy* 49 (3): 298–312.
Twain, Mark. 1994. *The Adventures of Huckleberry Finn. Penguin/Godfrey Cave Edition*. London; New York: Penguin Books. https://search.library.wisc.edu/catalog/999730567102121.
United Nations (General Assembly). 1966. International Covenant on Civil and Political Rights. *Treaty Series* 999 (December): 171.
Velleman, J. David. 1999. The Voice of Conscience. *Proceedings of the Aristotelian Society* 99 (1): 57–76.
Wolfe, Robert M., and Lisa K. Sharp. 2002. Anti-Vaccinationists Past and Present. *BMJ (Clinical Research Edition)* 325 (7361): 430–432. https://doi.org/10.1136/bmj.325.7361.430.

Chapter 6
Conscientious Objection in Healthcare: A Jewish Ethical Perspective on Balancing Professional Integrity and Patient Care

Jason Weiner

Abstract This chapter examines the role of conscientious objection in healthcare, highlighting its significance for practitioners who face ethical dilemmas in participating in certain medical interventions. The debate on conscientious objection often centers around issues of autonomy, moral integrity, and professional responsibility. This chapter offers a unique contribution by exploring the Jewish legal perspective, which frames conscientious objection through the lens of "placing a stumbling block before the blind." According to this interpretation, involvement in morally contentious actions is permissible if the act would proceed regardless of one's participation, thereby avoiding direct moral compromise. This approach provides an alternative framework for conscientious objection, allowing healthcare professionals to maintain both personal integrity and commitment to patient care. The discussion argues that religiously motivated ethical positions can offer practical insights for inclusive bioethics, and concomitantly contribute towards a healthcare environment that respects moral diversity while upholding equitable patient access.

6.1 Introduction

Although theologians were central to some of the earliest deliberations related to contemporary bioethics, in more recent years many theologians have become less interested in contributing to this public discourse, while many secular bioethicists have become less interested in input from religious perspectives. This development

J. Weiner (✉)
Spiritual Care Department, Cedars-Sinai Medical Center, Los Angeles, CA, USA
e-mail: jason.weiner@cshs.org

K. R. Smith (ed.), *Bioethics and Religion*, The International Library of Bioethics 116, https://doi.org/10.1007/978-3-032-33538-8_6

has been detrimental to each discipline.[1] It is essential that religious perspectives have a central seat at the table of bioethics, because such important issues must be settled by society as a whole, not by individuals or the scientific community alone.[2] Public debate and scrutiny are essential to educating the broader society and improving public confidence and consensus.[3] It is thus especially crucial to include religious communities in the conversation, because they can help develop broader buy-in and facilitate reflections related to meaning, values, and diverse perspectives.[4] Furthermore, theologians, and the Jewish tradition, have important perspectives that can contribute significantly to crucial contemporary health care deliberations. This is especially the case when it comes to the topic of conscientious objection, since religious concerns frequently underlie such objections.

6.2 Conscientious Objection

Some health care providers do not feel comfortable participating in certain interventions that go against their personal core values. Refusal to participate in these interventions is known as "conscientious objection," which has been defined as "the rejection of some action by a provider, primarily because the action would violate some deeply held moral or ethical value about right and wrong."[5]

The issue of conscientious objection is ancient, finding expression in biblical values.[6] It arose in modern medicine when some physicians refused to comply with abusive forms of medicine forced onto a population by totalitarian regimes, as was

[1] Michael McCarthy, Mary Homan, and Michael Rozier, "There's No Harm in Talking: Re-establishing the Relationship between Theological and Secular Bioethics," *American Journal of Bioethics* 20, no. 12 (2020): 12.

[2] W. Isaacson, *The Code Breaker: Jennifer Doudna, Gene Editing, and the Future of the Human Race* (New York: Simon & Schuster, 2021), pp. 331, 357–358.

[3] L.G. Locke, "The promise of CRISPR for human germline editing and the perils of 'playing God,'" *The CRISPR Journal* 3(1) (2020): 30.

[4] J. Lee, "The CRISPR revolution in genome engineering: perspectives from religious ethics," *J Relig Ethics* 50(3) (2022): 3–5.

[5] Jere Odell, Rahul Abhyankar, Amber Malcom, and Avril Rua, "Conscientious Objection in Health Professions: A Reader's Guide to the Ethical and Social Issues" (June 14, 2014), retrieved from https://scholarworks.iupui.edu/bitstream/handle/1805/4531/conscientiousobjectionscope.pdf?sequence=1andisAllowed=y.

[6] When King Saul ordered the commanders of his army to execute all of the innocent Kohanim of the city of Nov (I Sam. 22:17), they refused to carry out the king's orders because there was no duty to obey the king if his order was sinful (see *Radak* there; *Sanhedrin* 49a, *Rashi* s.v. *sheheim darshu*). Prior to that, the first recorded instance of civil disobedience is that of the midwives, Shifra and Puah, whose fear of God motivated them to refuse to obey the immoral order of the Egyptian king to kill the Jewish baby boys (Ex. 1:17–21); see discussion in Jonathan Sacks, *Covenant & Conversation, a Weekly Reading of the Jewish Bible: Exodus* (New Milford, CT: Maggid, 2010), 21.

the case in Nazi Germany.[7] It continues to be relevant in contemporary health care settings in terms of the participation of health care professionals—such as doctors, nurses, and pharmacists—in certain procedures, including euthanasia, medical aid in dying, contraception, and other reproductive technologies. Following the 1973 *Roe v. Wade* abortion decision, the United States Congress passed "conscience clause" legislation, known as the "Church Amendment," which protected an individual's right of refusal to take part in abortion and sterilization (42 U.S.C. 300a-7).[8] This amendment prohibited courts and government agencies from requiring individuals or facilities to perform abortions or sterilizations if they had moral objections, and protected individuals from employer discrimination if they were unwilling to perform abortions or sterilizations.[9]

Such legislation has not been without controversy, and several states, as well as many countries, have taken extremely divergent approaches.[10] The debate surrounding the appropriateness of health care providers engaging in conscientious objection has been intense and often polarizing.

6.3 Arguments in Favor of Conscientious Objection

While some argue that conscientious objection is a matter of freedom to practice one's religion, most feel that the foundation for exercising this right is even more fundamental and can be articulated in secular terms.[11] Many contend that a multicultural and tolerant society should simply not be attempting to impose ethical beliefs on others, but rather should strive to accept various moral viewpoints. They thus see respect for conscientious objection as a form of tolerance for "moral diversity."[12] Similarly, others argue that conscientious objection is a type of ethical

[7] Margaret W. Beal and Joyce Capiello, "Professional Right of Conscience," *Journal of Midwifery and Women's Health* 53 (2008): 406–12.

[8] *Roe v. Wade*, 410 US 113 (1973). The legal concept of "conscientious objection" dates back at least to the 1948 Universal Declaration of Human Rights. Article 18 states, "Everyone has the right to freedom of thought, conscience, and religion." Thaddeus Mason Pope, "Legal Briefing: Conscience Clauses and Conscientious Refusal," *Journal of Clinical Ethics* 21, no. 2 (2010): 163–80.

[9] Pope, 163–80.

[10] Christian Munthe, "Conscientious Refusal in Health Care: The Swedish Solution," *Journal of Medical Ethics* 43 (2017): 257–59; and Pope, "Legal Briefing," 163–80.

[11] Daniel P. Sulmasy, "Tolerance, Professional Judgment, and the Discretionary Space of the Physician," *Cambridge Quarterly of Healthcare Ethics* 26 (2017): 18–31; and Daniel Weinstock, "Conscientious Refusal and Health Professionals: Does Religion Make a Difference?," *Bioethics* 28, no. 1 (2014): 8–15.

[12] Mark R. Wicclair, "Conscientious Objection in Medicine," *Bioethics* 14, no. 3 (2000): 207; Stephen Wear, Susan Lagaipa, and Gerald Logue, "Toleration of Moral Diversity and the Conscientious Refusal by Physicians to Withdraw Life-Sustaining Treatment," *Journal of Medicine and Philosophy* 19 (1994): 147–59; and H. Tristram Engelhardt, *The Foundations of Bioethics* (New York: Oxford University Press, 1986).

modesty or humility, because it recognizes that one might be mistaken and should thus not be dogmatic regarding differing viewpoints.[13] To obligate one to go against their religious or ethical convictions is also regarded as a form of discrimination and a violation of ethical and human rights.[14] Some even see conscientious objection as being integral to a functioning democracy. They suggest that just as people should have the essential right to object to being drafted into military service, they should also have the right to refuse to engage in behaviors they see as immoral.[15]

However, what a professional can object to is limited, and some things cannot be tolerated, such as unnecessarily violating patient confidentiality or engaging in discrimination based on race, sexual orientation, and so on.[16] Some thus claim that it is the principle of respect for autonomy that is the strongest basis for both a patient's right to refuse treatments and a clinician's right not to engage in them.[17] Respect for individual autonomy implies that professionals should not have to abandon their morals in order to get a job, or to leave their job if new interventions become required of them, and that individuals should be allowed to select which procedures they are comfortable participating in as part of a right to free choice of employment.[18] Just as reproductive rights and abortion are often seen as matters of personal choice, so should a professional have a choice *not* to participate.[19]

Beyond autonomy arguments, many thinkers frame the importance of conscientious objection as most fundamentally respecting the need of individuals to maintain their "moral integrity."[20] This formulation allows health care providers to see themselves not only as technicians, like auto mechanics, by recognizing that someone who engages in procedures that go against their conscience and violate their

[13] Daniel P. Sulmasy, "What Is Conscience and Why Is Respect for It So Important?" *Theoretical Medicine and Bioethics* 29, no. 3 (2008): 144.

[14] Bernard M. Dickens and Rebecca J. Cook, "The Scope and Limits of Conscientious Objection," *International Journal of Gynecology and Obstetrics* 71, no. 1 (2000): 71–77.

[15] Julie Cantor and Ken Baum, "The Limits of Conscientious Objection—May Pharmacists Refuse to Fill Prescriptions for Emergency Contraception?" *New England Journal of Medicine* 351, no. 19 (November 4, 2004): 2008–12.

[16] Jason T. Eberl, "Conscientious Objection in Health Care," *Theoretical Medicine and Bioethics* 40 (2019): 483–86, https://doi.org/10.1007/s11017-019-09511-x.

[17] Kimberley Brownlee, *Conscience and Conviction: The Case for Civil Disobedience* (Oxford: Oxford University Press, 2012), 128–39, 167–68, 171–72.

[18] Bjørn K. Myskja and Morten Magelssen, "Conscientious Objection to Intentional Killing: An Argument for Toleration," *BMC Medical Ethics* 19, no. 82 (2018), https://doi.org/10.1186s12910-018-0323-0.

[19] Cantor and Baum, "The Limits of Conscientious Objection," 2008–12.

[20] Mark R. Wicclair, "Conscientious Objection, Moral Integrity, and Professional Obligations," *Perspectives in Biology and Medicine* 62, no. 3 (2019): 543–59; Martin Benjamin, "Conscience," in *Encyclopedia of Bioethics*, edited by Warren T. Reich, 3rd ed. (New York: Simon and Schuster Macmillan, 2004), 514; James F. Childress, "Appeals to Conscience," *Ethics* 89, no. 4 (1979): 315–35; James F. Childress, "Conscience and Conscientious Actions in the Context of MCOs," *Kennedy Institute of Ethics Journal* 7, no. 4 (1997): 403–11; and Jeffrey Blustein, "Doing What the Patient Orders: Maintaining Integrity in the Doctor-Patient Relationship," *Bioethics* 7, no. 4 (1993): 289–314.

own morality will feel that they are not being true to themselves.[21] Such moral distress can cause a loss of self-respect and may lead to burnout, fatigue, and emotional exhaustion.[22] In addition to such possible negative implications, some claim that a professional only becomes and remains a responsible moral agent through a gradual process of observation, reflection, and practice, which must be protected so that it can develop.[23]

From this perspective, protecting the moral integrity of health care professionals can benefit the entire society, since many people would want to have their own morality and integrity respected, and everyone benefits from having medical professionals who are trustworthy and honest moral public servants. Such behavior would not be promoted if professionals were asked to forsake their ethics in some situations, yet apply them to patient welfare in other areas.[24] Moreover, allowing students and residents conscience-based exemptions might increase their ethical sensitivity.[25] Protecting moral integrity and a health care professional's right to refuse to participate thus promotes the common good by enabling society to benefit from professionals who are encouraged to practice with these virtues.[26] Some also make the case that allowing practitioners to set limits on the interventions they are morally comfortable performing allows health care professionals to practice their vocation

---

[21] James F. Childress and Mark Siegler, "Metaphors and Models of Doctor-Patient Relationships: Their Implications for Autonomy," *Theoretical Medicine* 5, no. 1 (1984): 17–30; and Eva M. Norberg, Helge Skirbekk, and Morten Maggelson, "Conscientious Objection to Referrals for Abortion: Pragmatic Solution or Threat to Women's Rights?" *BMC Medical Ethics* 15, no. 15 (2014), https://doi.org/10.1186/1472-6939-15-15.

[22] Wicclair, "Conscientious Objection, Moral Integrity"; Vicki D. Lachman, "Conscientious Objection in Nursing: Definition and Criteria for Acceptance," *MedSurg Nursing* 23, no. 3 (2014): 196–98; and Lilia Susana Meltzer and Loucine Missak Huckabay, "Critical Care Nurses' Perceptions of Futile Care and Its Impact on Burnout," *American Journal of Critical Care* 13 (2004): 202–8.

[23] Mary Neal and Sara Fovargue, "Is Conscientious Objection Incompatible with Healthcare Professionalism?" New Bioethics 25, no. 3 (2019): 221–35.

[24] Charles D. Hepler, "Balancing Pharmacists' Conscientious Objections with Their Duty to Serve," *Journal of the American Pharmacists Association* 45, no. 4 (2005): 434.

[25] For example, medical students and residents have successfully objected to performing educational procedures on corpses without premortem consent, as well as the use of animals in medical schools, while other studies have shown that many medical students have felt pressured to act unethically, hindering their ethical growth. Lisa K. Hicks, Yulia Lin, David W. Robertson, Deborah L. Robinson, and Sarah I. Woodrow, "Understanding the Clinical Dilemmas That Shape Medical Students' Ethical Development: Questionnaire Survey and Focus Group Study," *British Medical Journal* 322, no. 7288 (March 24, 2001): 709. On premortem consent, see Mark W. Fourre, "The Performance of Procedures on the Recently Deceased," *Academic Emergency Medicine* 9, no. 6 (2002): 595. On the use of animals in medical schools, see Susan Dodge, "Under Pressure from Students, Medical Schools Offer Alternatives to the Use of Live Animal Experiments," *Chronicle of Higher Education*, November 15, 1989, A41, A43.

[26] Morten Magelssen, "When Should Conscientious Objection Be Accepted?" *Journal of Medical Ethics* 38 (2011): 18–21.

with individual integrity and thus earn their patients' trust.[27] While some will argue that the public good is ultimately not served by restricting access to health care, if other providers are willing to offer such services without excessive burden to the patient, a pluralistic society is better served by respecting the interests of as many individuals' rights as possible—both those in search of these services and those who do not want to engage in them.[28]

Principled opposition has arisen to the idea that medical professionals should have a right to conscientious objection, and the consensus of bioethicists is that some limits on conscientious objection may be appropriate.[29] One suggestion is that interventions for which conscientious objection may be requested should be limited, only allowing those with direct participation to object, as well as requiring objectors to satisfy certain tests of genuineness, reasonableness, or rationality before being permitted to refuse involvement.[30]

Broader support exists for requiring steps to ensure that conscientious objection has minimal negative impact on the quality, efficiency, and equitability of health care. To achieve that goal, many feel that health care providers should, at a minimum,

- offer emergency lifesaving services when no alternative is available[31];
- clearly inform patients and employers well in advance of what services they would not be willing to perform[32]; and
- never obstruct access to health care and ideally provide other options to receive those services, including respectfully providing information, options, or a timely referral.[33]

---

[27] Neal and Fovargue, "Is Conscientious Objection Incompatible with Healthcare Professionalism?" 221–35.

[28] Mark R. Wicclair, "Is Conscientious Objection Incompatible with a Physician's Professional Obligations?" Theoretical Medicine and Bioethics 29 (2008): 171–85. A related argument makes the point that just as medical professionals are sometimes free to refrain from offering services based on their personal interests or professional skills, they must also be allowed to refrain based on moral or religious objections. Aaron Ancell and Walter Sinnot-Armstrong, "How to Allow Conscientious Objection in Medicine While Protecting Patient Rights," *Cambridge Quarterly of Healthcare Ethics* 26, no. 1 (2017): 120–31.

[29] Neal and Fovargue, "Is Conscientious Objection Incompatible with Healthcare Professionalism?" 221–35; and Beal and Capiello, "Professional Right of Conscience," 406–12.

[30] Neal and Fovargue, "Is Conscientious Objection Incompatible with Healthcare Professionalism?" 221–35; Beal and Capiello, "Professional Right of Conscience," 406–12; and Eberl, "Protecting Reasonable Conscientious Refusals in Health Care."

[31] Lawrence Nelson, "Provider Conscientious Refusal of Abortion, Obstetrical Emergencies, and Criminal Homicide Law," *American Journal of Bioethics* 18, no. 7 (2018): 43–50.

[32] Anne Drapkin Lyerly, "Ethical Issues in Reproduction," in *Guidance for Healthcare Ethics Committees*, ed. D. Micah Hester and Toby Schonfeld (Cambridge: Cambridge University Press, 2012), 99; and Beal and Capiello, "Professional Right of Conscience," 406–12. See discussion in Beauchamp and Childress, *Principles of Biomedical Ethics*, 43–44.

[33] Lyerly, "Ethical Issues in Reproduction"; and Beal and Capiello, "Professional Right of Conscience."

Many conscientious objectors are unwilling to provide referrals for services that they morally object to, although some have found ways to make indirect referrals.[34] Some argue that when making referrals for procedures one is unwilling to perform, providers should pay for the increased costs their patients may incur, and others suggest that those unwilling to provide these minimal accommodations should be disciplined.[35]

6.4 Jewish Approach

This issue can become very heated and is often quite polarizing. While much of the above comes out of Christian—especially Catholic—opposition to engage in certain medical interventions, traditional Jewish law has many of the same concerns but offers a unique perspective that frames conscientious objection differently and can potentially provide a very helpful compromise approach.

The way conscientious objection becomes categorized in Jewish law takes the focus away from practitioners' own moral commitments and focuses on the results of one's actions. Jewish law forbids an observant Jew from transgressing the Torah, including some rabbinic prohibitions that serve as safeguards but are not required as strictly as biblical prohibitions. The questions Jewish law asks are: How concerned must one who observes Jewish law be when other people seek their support in doing something that violates Jewish law? and What biblical verse or category of rabbinic law does this fall under? Rabbinic authorities have placed the prohibition against taking part in another person's forbidden action under the category of the biblical prohibition against placing a stumbling block before the blind (Lev. 19:14). By categorizing conscientious objection this way, Jewish law can permit practitioners to do things like provide referrals and engage in ancillary support, such as anesthesiology, for most interventions, including those that biblical law prohibits, if the action is going to happen anyway.

This is because traditional rabbinic commentaries have understood this verse not only to be about actually tripping a blind person, but also to contain two fundamental prohibitions that take "blindness" more metaphorically:

1. This verse prohibits giving bad advice to a person who is "blind" in a certain matter or to the consequences of that advice (i.e., informational blindness).[36]

[34] Roger Trigg, "Conscientious Objection and 'Effective Referral,'" *Cambridge Quarterly of Healthcare Ethics* 26, no. 1 (2017): 32–43; and Norberg et al., "Conscientious Objection to Referrals for Abortion."

[35] Ancell and Sinnot-Armstrong, "How to Allow Conscientious Objection in Medicine"; and Savulescu, "Conscientious Objection in Medicine," 294.

[36] *Torat Kohanim* 2:14.

2. The verse also prohibits giving someone the means, causing, or facilitating an opportunity for them to stumble morally by transgressing the Torah.[37]

The Talmud and most medieval commentators focus primarily on this second, more expansive prohibition, which leads to much concern in Jewish law about the ramifications of one's actions and hence, the results-based categorization of conscientious objection in Jewish legal sources. The classic example in the Talmud is a prohibition against giving a cup of wine to a person who has vowed not to drink wine (a "*nazir*"), since giving them the cup can enable them to transgress their vow by drinking from it.[38]

While enabling sin is prohibited by this verse, this prohibition also has some crucial limits that are relevant for medical professionals. Most significantly, the Talmud rules that this commandment is violated only when the person extending the forbidden item and the person receiving it are located on two opposite sides of a river, which completely separates them, such that the person who is "morally stumbling" would be physically unable to access the forbidden item but for the assistance of the other person. Extending the forbidden item across the river places a "stumbling block" by creating the opportunity to sin and is thus prohibited by this verse.[39] However, one would not violate this verse by passing a cup to the person if both of them are on the same side of the river, meaning that the transgressor could have engaged in that prohibited action on their own without any help specifically from the one who passed the cup to them.[40] This ruling reconceptualizes the conscientious objection discussion by claiming that if a problematic action is going to occur anyway, then assisting with it is not as morally problematic.

Since enabling one to sin who could do so anyway without them (i.e., "same side of the river") does not transgress the narrow definition of the biblical prohibition, a medical professional would not be required by this biblical law to object to participating in any procedure that they might regard as immoral as long as the patient would still be able to receive it even without their involvement. However, the discussion does not end there, because the rabbis of the Talmud nevertheless prohibit that action under a category known as "*mesaye'a*" or "assisting" the sin of another.[41] This concept thus becomes the primary focus for many of those who analyze conscientious objection from a Jewish legal perspective: What actions are prohibited because they assist another person in engaging in a transgression of Jewish law, and which may one assist others with, even if they would not be comfortable with it for themselves? However, rabbinic authorities can maneuver this question with much

[37] *Pesachim* 22a; *Mo'ed Katan* 17a; and *Bava Metzia* 75b.

[38] *Avodah Zarah* 6b.

[39] *Avodah Zarah* 6b. See extensive discussion of this ruling in Rabbi Yair Hindin, "*Lifnei Iver*: When Best Practice Conflicts with Religious Practice," in *Sacred Training: A Halakhic Guidebook for Medical Students and Residents*, ed. Jerry Karp and Matthew Schaikewitz (New York: Ammud, 2018), 181–83.

[40] Hindin, 181–83.

[41] *Tosafot, Shabbat* 3a, s.v. *bava*; and Ritva and Ran, *Avodah Zarah* 6b, based on *Gittin* 61a.

flexibility, since it is categorized as a rabbinic prohibition and not a biblical one. This distinction lowers the degree of prohibition and allows numerous other leniencies to be considered, rendering such "assistance" permissible in certain circumstances.

For example, within this rabbinic prohibition against "assisting," the rabbis permit assisting the person who is assisting the transgressor, since the Torah forbids only placing a stumbling block before the blind, but not placing it before another person who will then place it before the blind.[42] This principle thus allows involvement in indirect actions. In these situations, where one can transgress without the assistance of another, many argue that the prohibition applies only if assistance is directly requested and provided at the precise moment of the transgression, but not if it takes place prior to the forbidden action.[43] This is a very relevant ruling when it comes to behaviors such as providing referrals or information. Some are also lenient if a prohibition is only violated by inaction, such as passively allowing prohibited actions to take place.[44] So, too, if one's livelihood depends on their professional duties, the rabbis do not require sacrificing that in order not to assist with a transgression that will be done anyway without them.[45]

The permissive category within these laws that may be the most relevant for medical professionals focuses on the claim that the reason the rabbis prohibited one from assisting another person's violation of Jewish law, even when that person can transgress on their own, is because the rabbis wanted people to discourage others from sinning. However, if the person who is sinning doesn't observe Jewish law, or knows exactly what they are doing and has decided to do the forbidden act without concern for any Torah prohibition that might exist, the rabbis did not require one to try to prevent them from doing so. Therefore, this rabbinic prohibition against assisting them doesn't apply.[46] This is especially true if allowing a person to engage in a prohibited act might prevent that person from violating an even worse prohibition instead. Similarly, if one's assistance allows others to violate a lesser prohibition, one is then required to consider the long-term consequences and favor actions that will lead to less overall sin.[47] This reasoning likely applies to most situations of

---

[42] *Avodah Zarah* 14a.

[43] Mishnah, *Avodah Zarah* 4:9; Responsa *Binyan Tziyon* 1:15; Responsa *Mashiv Davar* 2:32 (he adds that this would be prohibited only if one causes the action to happen sooner than it would have; otherwise, it is never prohibited to assist); and Responsa *Minchat Asher* 2:28, 30.

[44] Responsa *Melamed LeHo'il* 1:34.

[45] Ritva, *Gittin* 61a, s.v. *mashelet ishah*; Responsa *Mashiv Davar* 2:32; *Tzitz Eliezer* 19:33:9; *Shiurei Torah LeRofim* 4:246(2), 269; and Y. Schechter, *Kuntres Harofeh LeShevurei Lev: Psychotherapy in Halachah*, 72, 74.

[46] *Iggerot Moshe*, YD 1:72 explaining *Shach*, YD 151:6 and *Dagul MeRevavah*, YD 151:6. See detailed discussions of this ruling in *Masorat Moshe*, 4:321–22; *Tzitz Eliezer* 14:85(3); Hindin, "*Lifnei Iver*," 184–85; and Yitzchak Schechter, *Kuntres Harofeh LeShevurei Lev: Psychotherapy in Halachah*, (New York: printed by the author, 2021), 67. See also *Yabia Omer* 9: OC108(190). R. Elyashiv also ruled like R Feinstein on this matter (*Nishmat Avraham OH* pg. 829).

[47] *Iggerot Moshe*, YD 1:72. Rabbi Feinstein argues that even those who disagree with his approach would agree in the case of permitting renting a hall to Jews who will use it for Jewishly inappropri-

conscientious objection today, when patients turn to medical professionals for a desired intervention, but not their moral perspective on the matter.

While not all contemporary rabbinic authorities accept this approach, many do, and its ramifications for the conscientious objection discussion are profound.[48] Although these principles arise specifically out of Jewish law, perhaps this "two sides of the river" paradigm discussed above provides a different way to look at religious objection in general, which can be incorporated by many health care professionals. Following this approach would mean that a medical professional could, or perhaps *should*, be involved in most aspects of patient care for a patient who has decided to engage in an action that the medical professional would otherwise object to, since if one professional does not do it, others will. Jewish law thus permits a medical professional to be involved in any service or procedure that the patient will be able to receive anyway, without their participation, if their assistance is indirect and does not directly violate Jewish law at the moment any transgression occurs.

Framed this way, one need not see such assistance as a compromise of their values, but rather a recognition of the fact that people have options. Not only is voicing conscientious objections not always effective, but it could also have counterproductive consequences and lead to more sinning, rather than less. For example, research related to physician aid in dying has shown that, paradoxically, a nonjudgmental, supportive approach from clergy has been more effective in allowing patients to consider alternatives to aid in dying, and even to ultimately change their minds, than active opposition to the patient's decision.[49] Studies have also shown that when clinicians take the time to listen compassionately to their patients and explore the reason(s) for their requests, that act of simply listening to the patients' concerns

---

ate activities, since the alternative is for them simply to rent another hall where they won't serve kosher food, thereby compounding their transgression of Jewish law. Furthermore, Rabbi S. Z. Auerbach writes in Responsa *Minchat Shlomo* 1:35(1) that one may give food to a nonobservant Jew who will not cover their head or recite the required blessings, since forcing them to do so could cause people to hate the Torah or religious Jews. This ruling is based on Rabbi A. Eiger's commentary on *Yoreh De'ah* 181:6 that since the Torah prohibits a man from shaving his face with a razor but doesn't prohibit a woman from doing so, it would be better for a man who is going to shave his face with a razor anyway to have a woman do it for him, even though she is thus violating the prohibition of placing a stumbling block (or only the rabbinic prohibition of "assisting," since he can do it himself). Rabbi A. Eiger argues that when an action is prohibited by the Torah only due to placing a stumbling block in front of another, then it is better for a person to perform such an action, since this way ultimately fewer sins will be committed. See also *Shiurei Torah LeRofim* 6:478, 597, which says that whenever one's actions will ultimately lead to long-term positive results, it is not considered a "stumbling block"; see also explanation of Rabbi A. Lichtenstein on this point in Hindin, "*Lifnei Iver*" 176.

[48] For those who do not accept this approach, see Responsa *Minchat Shlomo* 3:103(4); Steinberg, *HaRefuah KeHalakhah* 2:380; and *Nishmat Avraham*, OC 656:1(3) (820–21 in 3rd ed.); see also Rosman, *Petichat HaIggeret*, 356. For those who do accept this approach, see Responsa *Maharit* 1:97; Responsa *Minchat Asher* 2:28–30 and vol. 4 (forthcoming).

[49] Bryant Carlson, Nicole Simopolous, Elizabeth R. Goy, Ann Jackson, and Linda Ganzini, "Oregon Hospice Chaplains' Experiences with Patients Requesting Physician-Assisted Suicide," *Journal of Palliative Medicine* 8, no. 6 (2005): 1165; and Linda Ganzini and Steven K. Dobscha, "If It Isn't Depression…," *Journal of Palliative Medicine* 6, no. 6 (2003): 927–30.

helps to mitigate many of them.[50] Similarly, regarding abortion, many women who don't have access to legal services end up attempting to abort the fetus themselves in very dangerous and even fatal ways.[51]

Therefore, as long as medical professionals are not directly performing the prohibited interventions themselves, Jewish law often permits their involvement with what could be considered "immoral" interventions, such as:

- writing prescriptions for prohibited medications[52];
- sharing information/describing options, even if some may be prohibited by Jewish law, with a patient or making direct and effective referrals to other providers, even if they may carry out actions that Jewish law would prohibit[53];
- serving on a board that permits abortions or other such procedures[54];
- writing DNR orders[55];
- enrolling and caring for a patient in hospice[56];
- administering anesthesia or other types of supportive care for abortion or other controversial procedures[57]; and
- providing marital therapy for a couple in a union forbidden by Jewish law.[58]

---

[50] Dale A. Matthews, Anthony L. Suchman, and William T. Branch Jr., "Making 'Connexions': Enhancing the Therapeutic Potential of Patient-Clinician Relationships," *Annals of Internal Medicine* 118, no. 12 (1993): 973–77.

[51] Lisa Rosenbaum, "Perilous Politics—Morbidity and Mortality in the Post-Roe Era," *New England Journal of Medicine* 381 (2019): 893–95; and David A. Grimes, Janie Benson, Susheela Singh, Mariana Romero, Bela Ganatra, Friday E. Okonofua, and Iqbal H. Shah, "Unsafe Abortion: The Preventable Pandemic," *Lancet* 368 (2006): 1908–19.

[52] *Nishmat Avraham* 3:5 (159 in 3rd ed.); see also Hindin, "*Lifnei Iver*," 179.

[53] Responsa *Minchat Asher*, vol. 4 (forthcoming); see also Hindin, "*Lifnei Iver*," 186. It is important to point out that there is also room for leniency when there is a likelihood that even if facilitated, the sin will not end up occurring. Responsa *Shevet HaLevi*, YD 62 based on *Shulchan Arukh*, YD 151:1.

[54] Responsa *Minchat Asher*, vol. 4 (forthcoming) and 2:30(6). On the other hand, Rabbi Zilberstein writes that when such a committee allows an abortion that would not be permitted by Jewish law, an observant Jew must either excuse themselves from that committee or clearly protest the decision. *Shiurei Torah LeRofim* 6:464, 559.

[55] *Nishmat Avraham*, YD 339:4(7) (510 in 3rd ed.). Rabbi M. Feinstein also explains why this would be permitted in many scenarios, but he suggests removing oneself from the situation if possible and prohibits it in certain scenarios (*Iggerot Moshe*, YD 4:54).

[56] *Nishmat Avraham*, YD 339:1(4) (499 in 3rd ed.).

[57] Responsa *Minchat Asher* (forthcoming; see quotation "Becoming a Physician: When Secular Law and Values Clash with Halacha," https://e8b7a5f9-e27e-4732-8c9b-1f83787ae640.filesusr.com/ugd/9c4394_7c5cdd45e1774f0c96fdbb00a3552fc9.pdf); *Tzitz Eliezer* 19:33(1). On the other hand, Rabbi Moshe Feinstein is quoted as ruling that it is forbidden to serve as an anesthesiologist for an abortion procedure (*Masorat Moshe*, 4:428). See extensive discussion on serving as an anesthesiologist for abortion or as a surgeon for a forbidden procedure in *Nishmat Avraham OH* pg. 815–24.

[58] Y. Schechter, *Kuntres Harofeh LeShevurei Lev: Psychotherapy in Halachah*, 59, in the name of Rabbi Dovid Cohen (see extensive analysis of the topic there).

6.5 Conclusion

While some secular bioethicists prefer not to have religious practitioners or theologians at the table because of assumptions about their irrationality or unreasonable requests, it seems to me that the perspectives detailed above are ethically defensible, logical, and fair. In particular, I have outlined a perspective that teaches that one does not place a stumbling block when their assistance does not directly induce more sin. Indeed, objecting can become an exercise in futility or potentially cause even worse outcomes than might ensue by refusing to participate. In particular, when the individual planning to commit an act that the professional sees as immoral will not be swayed by the protest, and the act can be carried out in any event without that individual provider, we can invoke the "same side of the river" paradigm that arises out of Talmudic law. Rather than focusing specifically on the practitioner's own conscience, a practitioner can then justify taking part in most interventions. In this way, one can maintain their moral integrity even without engaging in conscientious objection, while resting assured that they have not actually placed a stumbling block before the blind.

It remains important to protect moral diversity and practitioners' integrity, and to allow conscientious objection in some cases, but it need not be invoked nearly as frequently as it currently is. Those who believe that it is important for religious individuals, or anyone with a strong set of moral principles, to be fully involved in health care so that they can take culturally and religiously sensitive care of patients and have a positive impact on the way medical care is seen and provided in our society, must find ways of working within the system.[59] Otherwise, they might be forced to simply leave some of the most crucial interventions only to those who do not share their values.

[59] Some argue that a health care provider should not impose their own religious commitments onto specific patients if doing so is likely to turn that patient off from seeking essential health care. Rabbi N. Bar-Ilan, quoted in Hoffman, "*Dilemot vekonfliktim hilkhati'im umusari'im bahem nitkalim psichologim dati'im,*" November 11, 2008, https://www.hebpsy.net/articles.asp?id=1896), particularly because health care providers engaging in their professional duties does not imply formal religious approval of their patients' actions and should therefore not be seen as "assisting" their sin. Rabbi S. Rappaport, https://www.hebpsy.net/articles.asp?id=1896, November 11, 2008. I thank Rabbi M. Torczyner for pointing these two sources out to me. On the importance of ensuring that there are religious doctors, see *Chelkat Yaakov*, YD 190(1).

Chapter 7
Should Religion Be Permitted a Seat at the Table of Bioethics? The Case of Abortion

Alberto Giubilini

Abstract In addressing the question whether religion should be permitted a seat at the table of bioethics, and specifically where abortion is discussed, I argue that the question requires first some disambiguation. There is a sense in which the answer is obviously affirmative, but also a sense in which the answer is negative. In particular, when it comes to bioethical discussion about the appropriate scope of the healthcare profession and about policy making, I argue that religion itself is neither sufficient nor necessary for deliberating about abortion in liberal societies. However, even if its role is neither sufficient nor necessary, religion can still have an important, albeit merely instrumental, role to play in that kind of deliberation. Its values and norms can be grounded in non-religious views, such as natural law theory, that are relevant indeed to bioethical discussion on abortion. Thus, religion—and particularly Christianity—can be one route into that kind of bioethical discussion for many people. However, while these values and norms might well be valid and widely supported, they are also open to public scrutiny, political negotiation, and, in some cases, defeat. Those supporting the role of religion in bioethical discussion on abortion would need to accept this implication.

7.1 Introduction

The question whether religion should have a seat at the table of bioethics is, in one form or another, frequently asked. It is also too vague to be given a straightforward 'yes' or 'no' answer. Not that this vagueness has prevented some scholars from confidently providing one (see e.g. Biggar 2015 for a confident 'yes' and Murphy

With many thanks to Christina Lamb and Ryan Kulesa for very helpful feedback.

A. Giubilini (✉)
Senior Research Fellow, Uehiro Oxford Institute, University of Oxford, Oxford, UK
e-mail: alberto.giubilini@philosophy.ox.ac.uk

K. R. Smith (ed.), *Bioethics and Religion*, The International Library of Bioethics 116, https://doi.org/10.1007/978-3-032-33538-8_7

2012 for a confident 'no'). However, a straightforward answer to a vague question will inevitably be vague itself. For what is the table of bioethics? Is it the table where the morality of certain medical procedures is discussed? Or the one where the boundaries of appropriate healthcare or medicine as professional settings are drawn? Or where policy decisions on bioethical issues are made?

Depending on which question we choose, the answer might differ. Thus, it would be good to clarify which specific question one is trying to answer, before offering a confident 'yes' or 'no', or indeed any answer at all.

In this chapter, I will discuss three questions regarding abortion specifically, namely whether religious values should have a place at the table:

1. where the morality of abortion is discussed
2. where it is discussed whether abortion is part of appropriate healthcare and medicine
3. where abortion legislations are discussed

However, before doing that, some terminological tidying up is in order.

7.2 Terminological Tidying up: On Bioethics, Medicine, Healthcare, and Religion

A recent special issue of the journal *Bioethics* was dedicated to discussing the role of religious pluralism in *healthcare* (see Oakley et al. 2023). A few years earlier, the *Journal of Medical Ethics* hosted a debate between Nigel Biggar and his critics on whether religion deserves a place in 'secular *medicine*'. And the volume in which this chapter appears is about the place of religion in *bioethics*. These three terms—healthcare, medicine, and bioethics—refer to different things. The answer to a question about the relationship between religion and bioethics is not necessarily the same as the answer to a question about the relationship between religion and (secular) medicine or healthcare.

Bioethics is the field where values and norms informing both medicine and healthcare are discussed. For instance, is abortion morally permissible and, if so, when? When, if ever, is medical assistance in dying morally permissible, and in what form? Who has stronger moral claim to priority access to scarce medical resources? And so on. The question whether religion should be permitted a seat at the table of bioethics *so understood* has a straightforward, but perhaps not too interesting, answer: yes. Religion typically contains values and norms that are relevant to healthcare and medicine, and this is precisely what bioethics is about. In this respect, religion is not different from secular, atheist, or agnostic views that come with their baggage of ethical values about health, life, and death.

There will be disagreement at that table, of course. That is precisely why the table for moral discussion of medical options exists. We sit there to discuss from our different moral, religious, and cultural backgrounds. The question whether religion should be permitted a seat at the table seems to suggest that someone has the authority to issue invitations, but who is in the position to issue invitations? There are no hosts and guests at the table of bioethics. Religious and non-religious people alike come equipped with their own of moral values that are relevant to the moral issues raised by medicine and healthcare. This is true regardless of whether we think of bioethics as an academic discipline or topical for a wider societal discussion. The first interpretation of the question mentioned earlier—whether religion should have a seat at the table where the morality of certain medical procedures, such as abortion, is discussed—is a question *about* bioethics itself.

A different, and perhaps more interesting, question is whether those values and norms that religion contributes to bioethical discussion can provide right, reasonable, or helpful answers to inform medicine, the healthcare profession, and policy making. For example, whether abortion provision should be part of the healthcare profession or a legally available option; whether access to abortion, if a right all, is a negative or a positive right; and so on. These are *bioethical* questions about professionalism, healthcare, and healthcare policy, not questions *about* bioethics itself.

Given the different issues at stake, it is important not to automatically turn an obvious answer to the first question into an answer to the other two. Unfortunately, that is what sometimes happens when authors give confident answers to questions whose terminology they have not clarified.

Here are two examples.

Timothy Murphy argues that religion should be excluded from "public discourse on bioethical issues" and replaced with an "irreligious bioethics". In his view, "religions ultimately rely on assumptions and claims that evade secular evaluation because they are typically unfalsifiable, infinitely mutable in the face of objections, rooted in personal experiences that defy independent analysis, or rooted in the murk of human history" (Murphy 2012).

But—one might be tempted to reply—so what? Let's leave aside for the moment that many would disagree with this way of characterizing religion (more on this later). Even if that was a correct view, at best this would be a reason to exclude religion from defining the boundaries of healthcare as a profession (second question) or from informing public policies around bioethical issues (third question).

It would not be a reason for excluding it from a discussion about the morality of certain procedures, which is what bioethics is about. In fact, there is a prior meta-ethical question to be asked about what exactly falls within the realm of bioethics and medical ethics, that is, what counts as ethics or morality (I will use these two terms interchangeably here). Whether religion should be part of bioethics in this sense depends on how that question is answered, and the answer would be terribly complicated. But that is no excuse for skipping it and moving to a conclusion about the exclusion of religion from that realm. Moreover, we may ask, what questions does religion seek to answer? And what questions does bioethics seek to answer? Are they mutually exclusive? It is hard to justify excluding religion from the table of bioethics without such preliminary discussion.

After all, assumptions and claims of religions contain moral values and norms that are relevant to the healthcare and medical decisions of many people. According to the last Census (ONS 2021), for instance, in England and Wales nearly half of the population (46.2%, or 27.5 million people) identifies as Christian and a significant minority (6.5% or 3.9 million) as Muslim. An "irreligious bioethics" would fail to provide some answers that are relevant to the healthcare choices of all of these people.

Conversely, Nigel Biggar (2015) thinks that "religion deserves a seat at the secular table of negotiation" in medicine, at least as long as the religious believer abandons "any sheer appeal to religious authority".

One reason is that in religion "there is truth—be it sometimes complex and internally plural—to be understood: as the creation of the one wise God, the world possesses a given rationality that is there for rational beings to grasp" (Biggar 2015, p. 231).

A truth that is there 'as the creation of the one wise God' is certainly relevant to Christians. This, again, supports an affirmative answer to the first question, i.e. whether religion deserves a seat at the table where the morality of certain medical and healthcare options is discussed.

But what about the second and third question, i.e. whether religion should have a say in defining the proper scope of the medical and healthcare profession or in shaping related policies? It is unclear whether Biggar means to extend his answer to those questions, but that seems to be implied by the idea of a 'table of negotiation'. What could the negotiation be about? Biggar does not clarify, but if it is about informing healthcare professionals' responsibilities or laws around, say abortion, that affirmative answer would be problematic.

According to the same Census, the second largest group in England and Wales is the "non-religious" (37.6% or 22.2 million people). Using the morality that Christians derive from "the one wise God" to design public policies and professional standards meant to apply to the non-religious would, by itself, take a lot out of the idea of 'liberal' society. A religious morality might be necessary both in society at large and within healthcare—a question I am happy to leave open. However, it would not be sufficient to preserve the liberal character of a society. A Jacobin exclusion of any God-inspired morality from informing healthcare professionalism and policies is inconsistent with liberalism, but so is the adoption of such morality as the standard for professionalism and policy making. If religious morality has to play a role in these matters, liberalism requires that it be integrated and compatible with additional reasons that are "in some sense, justifiable or acceptable to all those persons over whom the rules purport to have authority" (Quong 2022). That is the Rawlsian idea of 'public reason'. Biggar does attempt to provide such reasons, which I will discuss below to bring up what I take to be the main problems with them. For the moment, I simply want to point out that Biggar's positive answer remains vague. It might be an easy answer to the question about the role of religion in discussion of the *morality* of healthcare practices, but it is unclear whether he wants it to also be an answer to the more challenging questions on professionalism and policies.

The term "religion" also requires some clarification. What counts as 'religion' is contested. There is no way I can do justice to the complexity of the issue here. Buddhism, which is considered by most a religion, is typically brought up as the example that muddies the waters. It shows that a belief in a transcendent God is not a necessary condition for a worldview to count as religious. The presence of authorities or authoritative texts does not seem a sufficient or necessary condition, either. It is not sufficient, because some worldviews with authoritative texts like Marxism are not religions (Camosy 2012). And it is not necessary, because Buddhism, once again, does not have an authoritative book that is equivalent to the Bible or the Quran, but different canons in different cultural contexts. And of course, this is all contested. Some would not consider Buddhism a religion and some would consider Marxism a religion (Ling 1980).

Be that as it may, lack of a shared understanding of religion often means that contenders in this debate talk past each other. Religious scholars often accuse their opponents of arguing against a strawman by depicting religion as irrational and authoritarian, in opposition to a more rational and unbiased secular approach. That is, they accuse their opponents both of not providing an adequate definition of religion and, as a consequence, of failing to see that secular views are also based on assumptions not amenable to reason or evidence.

Indeed, some religious scholars argue that secular moral theories are ultimately based on *faith*. For example, Catholic theologian Charles Camosy talks of the "authoritative, transcendental, faith-based, first principles of traditions like hedonistic and preference utilitarianism" (Camosy 2012). I am not sure this accusation hits the mark. On one understanding, it seems to assimilate faith to moral intuition, which would fail to capture many aspects that we usually attribute to religious faith. For example, "[i]n the Christian context, faith is understood both as a gift of God and also as requiring a human response of assent and trust, so that their faith is something with respect to which people are both receptive and active" (Bishop and McKaughan 2023). This doesn't seem a prerogative of, say, the intuition that happiness is all that ultimately matters. However, Camosy's claim might be understood as the claim that, whatever moral view we endorse (religious or not), it involves faith in our own moral faculties (i.e., intuition and reason), that is, it requires at some point a response of assent and trust in those moral faculties. On this interpretation, it is not clear how faith in the religious sense is more problematic than faith in this secular sense.

In any case, if the question we are here considering has to be meaningful, we need at the very least a working definition of religion. I take it that a view that does not include any faith in God, or any appeal to some authority, or any sacred text does not count as religious, at least for the purpose of this discussion. I take these to be at least individually necessary, and the three of these jointly sufficient conditions for some view to count as religion, at least for the purpose of this discussion.

It is only against such understanding that the initial question whether religion deserves a seat at the table of bioethics makes sense. If there is no difference between what we commonly consider religion and other worldviews that we commonly do not consider religious, then there is no point in asking the question. For instance, if we say that something like Christianity and something like utilitarianism are equivalent in all the aspects (whatever they are) that identify something as a religion, we would be making the concept of religion rather meaningless for most communicative purposes—including the ones that lead to ask whether religion deserves a seat at the table of bioethics or of secular medicine.

I am now in the position to reframe our question as follows: can religion understood as a worldview based on faith and appeals to some authority (a person, a book, an institution) offer right, reasonable, or helpful answers to bioethical questions about abortion that can benefit general ethical discussion, professional conduct, and policy making? My answer is that it can, but not in ways that those defending the role of religion in bioethics are likely to support.

7.3 Abortion and Ethics: The Contribution of Religion

In the Platonic dialogue *Eutyphro*, Socrates asks "whether the pious or holy is beloved by the gods because it is holy, or holy because it is beloved by the gods". The dilemma is often employed to raise a more general problem about the relationship between morality and divine command. In that context, the Socratic question is restated as: Is something good because God commands it, or does God command something because that is good? The first option is problematic because it seems to imply that God's will, since it doesn't rely on any other criterion, is arbitrary. This is inconsistent with the idea of a perfect God. The second option is problematic because it seems to imply that God himself needs to rely on a source of morality, or rationality, that exists prior and independently of himself. This is inconsistent with the idea that God is the source of morality. All in all, the dilemma is taken to support the idea that morality cannot derive from God (Rachels 1995).

Within the debate on the relationship between bioethics and religion, a version of the dilemma is faced by those who claim that God's wisdom provides the world with a 'rationality that is there for rational beings to grasp', as Biggar puts it. On this view, that rationality and morality is what religion can bring to the table of bioethical discussion on abortion (Biggar 2015, 232). What is the source of that universally accessible rationality and rational morality that God provides?

Suppose that abortion contravenes the rational order of the world and is therefore objectively unethical. And suppose we agree that, if there is such a thing as an objectively unethical practice, the healthcare profession should not include it. For instance, if killing healthy newborns is objectively ethically unacceptable, then the healthcare profession should not include infanticide. Many, especially among religious people, would think of abortion in the same or a similar way. But what is religion's contribution to the claim that infanticide or abortion contravene the rationality of the world order and are therefore immoral?

One option is that rationality and morality are themselves "the creation of the one wise God'. This corresponds to the first horn of the Eutyphro dilemma. If God's will is the only source of this rationality or morality, then what we call 'rationality' or 'morality' is actually arbitrariness. One might well have faith in the goodness of this arbitrariness or in the relevant religious authority's pronouncement on it. However, when it comes to discussing the scope of a profession supposed to provide societally valuable services—like medicine and healthcare—or healthcare legislation in liberal societies, faith and religious authority are *not sufficient* (and arguably, not necessary either, as we shall see later) as sources of rationality, values, or norms. Lacking faith in God, there is no moral reason that is accessible to a non-religious person which could justify a certain norm and make it non-arbitrary to them. To the extent that a society wants to remain liberal, justifications for policies need to be independent of *mere* appeals to constitutive aspects of religion as defined above, such as faith or authority. As we have seen, Biggar himself says that religion does have a place in secular medicine only as long as it "abandons appeals to sheer authority". The problem is that, if we abandon appeals to faith and authority, we would need an independent, non-religious basis for the regulation of healthcare and medicine and the place of abortion in them.

Which leads to the second horn of the dilemma, namely that the standard for rationality and morality that religion can convey is not arbitrary, but accessible to rational beings regardless of faith or reliance on religious authority. That is what most theologians and religious scholars would endorse, and indeed it is a more promising basis for addressing the second and third question above. However, it does not require religion. The standard for rationality and morality exists prior to and independently of religion. This does not rule out that religion could still be valuable in discussion about the proper scope of healthcare and medicine or about legislations—after all, it is a way of conveying meanings and values that derive from a rational understanding of the world. But that would imply religion is valuable merely instrumentally.

Take the claims that abortion is morally impermissible, that it should not be part of healthcare delivery, and that it should be legally prohibited. It might well be that religious and non-religious reasons converge on these conclusions. But if so, what is religion's distinctive contribution to those moral claims?

Natural law theory is the route through which religious and non-religious views against abortion typically converge. In particular, natural law theory is central to Christianity. However, it is not a theological view and does not require religion to be understood, appreciated, and endorsed. Instead, "natural law ethical theory aids us in understanding which human actions are morally right or wrong through the aid of human reason alone—without the use of Divine Revelation or the teachings of the Magisterium" (Irving 2000, p. 47). It is based on a teleological, ultimately Aristotelian view of the natural world. The natural world possesses an intrinsic rationality, that is, it is governed by laws of nature that can be grasped through reason. The ontological claim—about the nature of the world—and the epistemological claim—about what we can know of it—are aligned because the rationality that characterizes nature is the same one that characterizes human capacity for reason.

In this rational order, each organism and parts thereof fulfil certain functions determined by their rational nature. These functions define both what is good and bad for these parts and what is right and wrong to do to them. Two aspects of it are relevant to an ethics of abortion. First, the telos, or one of the proper functions of the female human body, is reproduction, which abortion prevents. Second, the value or sacredness (depending on whether we are using secular or religious terminology) of being human lies in the rational nature we all share *qua* humans. That includes adult human beings and foetuses alike. It is precisely our capacity for reason that identifies us as human, even if in some of us—including foetuses—it is present merely in potency.

Potency is not an accidental property of the substance, but an essential property that defines its nature. The terminology here expresses the Aristotelian metaphysics that conceives of entities as consisting of matter (act) and form (potency)—or "hylomorphism". Potency is what defines what an entity is. Some of the strongest arguments from potentiality against abortion—i.e. arguments to the effect that the moral status of foetuses lies in their potential for becoming fully developed humans—are grounded in natural law theory. The potential to develop rational capacities is valuable because it implies that rationality is already present, even if merely in potency, in the very nature of human life (Reichlin 1997). Human beings at all stages of development, including before birth, have moral status that makes killing them morally impermissible before any mental capacity or capacity for perception is present in its actuality.

Now, let's ask the same question again: what is religion contributing to this debate? Natural law theory and potentiality arguments do not need to appeal to faith or religious authority. Rather, as per the second horn of the Eutyphro dilemma, they provide an ethical standard that precedes and is independent of religion and which religion itself can adopt. In fact, it is this Aristotelian metaphysics that has eventually been incorporated into the Christian doctrine, mostly through the Thomistic tradition. It is consistent with, and indeed supports, Christian views on ensoulment, where the soul is the reification of that potency. But the underlying metaphysics is not originally Christian and does not require faith or appeals to religious authority. Discussing the first horn of the Eutyphro dilemma, we said that religion is *not sufficient* as a source of rationality, values, and norms in liberal societies. This discussion suggests that religion is *not necessary*, either.

Interestingly, different interpretations of the application of natural law theory are also the source of some disagreement within the Christian tradition about the doctrine of 'delayed ensoulment'. This is an old Catholic and Thomistic view with some support from passages of the Old Testament (Dunstan 1988). According to it, ensoulment occurs some time after fertilization, rather than *at* fertilization. Even within that debate, disagreements and solutions at the scholarly level are not necessarily phrased in terms of faith or religious authority. Instead, scholars often debate around the secular counterpart of the same issue to support theological points. The secular counterpart concerns biological understandings of embryo development and philosophical understandings of potentiality, rather than the theological construct of the soul. According to Norman Ford—prominent theologian, Catholic priest, and Salesian friar—it is only when the embryonic primitive streak is formed, and twinning is no longer possible that "a living individual human body is actually formed with the active potential to develop further without change in ontological identity" (Ford 1988). Ford is developing a philosophical argument in support of the idea that the relevant type of potentiality of foetuses, and therefore their full moral status, only starts to exist about 2 weeks after fertilization, when meaningful and ethically relevant individuality can be attributed to it. Other Christian thinkers, however, argue that a proper understanding of embryology and of the concept of potentiality implies that individuality, and therefore a human life in the morally relevant sense, is present from the moment of fertilization (Tonti-Filippini 1989).

This is a very sophisticated debate that I don't have the space and the level of knowledge to adequately represent here. For the purpose of this discussion, I just want to point out that this kind of disagreement concerns reasons that are not based on faith, appeal to religious authority or sacred texts. They require reasoning on the concept of potentiality and embryo development.

As Diane Irving puts it, "because the basic precepts of natural law theory are proximately grounded on an objectively knowable human nature, they are applicable to all human beings, precisely because we all possess such human natures. The possession of natures which are specifically human is precisely what we all have in common" (Irving 2000, p. 48). And this is exactly what, on this view, makes the life of a foetus and the life of an adult human being equally valuable: they share the same nature. Abortion violates objective moral obligations that are determined by the laws of nature. This is where religious and non-religious views against abortion can converge. That is, by adopting a rational and ethical standard that is non-religious in nature, as per the second horn of the Eutyphro dilemma applied to bioethics.

So if we do accept that what distinguishes religion from other types of worldview is faith in a God, or any religious authority, or any sacred text, or any combination thereof, then religion might well deserve a seat at the table of bioethics when abortion policies or the place of abortion in healthcare are discussed. However, on the assumption that we want to live in a certain type of liberal society, it does so only as long as it leaves out what distinguishes it as religious, that is, faith or appeals to religious authorities.

7.4 Abortion and Public Discussion: The Contribution of Religion

Another type of contribution that religion is sometimes said to make to bioethical discussion is by providing what we might call 'procedural' values. These are values that do not inform substantive bioethical positions such as those in favour or against abortion. Instead, they are values that can promote a respectful, open, and pluralistic discussion about controversial policies on bioethical matters. Biggar for example mentions the 'conversational virtues' that Christianity can foster. These include a sense of finitude and fallibility that encourages an "openness to be taught and corrected"; an awareness of being sinners which promotes a "readiness to confess conversational dishonesty"; a cautiousness in identifying the true with the familiar, which opens to tolerance of unwelcome views; patience with frustration; and "forgiveness as a reaction to conversational injustices such as misrepresentation by others of one's own views", which Biggar thinks is precisely what secular thinkers do when they associate Christianity with authoritarianism and irrationality. This is all unobjectionable. If there is some arrogance in secular thinkers that Christianity or some other religion can correct, that would be a very helpful contribution indeed to bioethics, secular medicine, healthcare.

So far so good. However, there is more that religion is sometimes said to bring to bioethical discussion in terms of procedural values. This is a certain idea of equality of individual human beings. That is where the problem arises with bringing religion into discussion of abortion policies and of the place of abortion in professional healthcare and medicine. As is often the case, the language of 'equality' is fishy, ambiguous, and at risk of being abused. When it comes to discussing abortion, religion contributes to this ambiguity.

According to Roger Trigg, "[f]rom a historical perspective, ideas of equality certainly grew on Christian soil. The idea that we are all equal, in the sight of God who created us, has been powerful. From God's point of view, it has been claimed, all differences of rank, social position, and so on are irrelevant. We are all God's children" (Trigg 2007, 82). He claims that such idea spread from its original Christian roots to inform charts of rights, legislations, and international treaties across liberal countries. All other fundamental values that Western societies hold dear, including freedom, liberties, and justice, derive from this more fundamentally Christian idea of equality among individuals.

This notion of equality is taken to ground a principle of toleration. That is, using the terminology adopted above, equality is understood as a procedural value that can contribute to accepting diverse views in bioethical discussion on controversial issues. Triggs refers to John Locke's justification of toleration in society at large. That justification is fundamentally Christian in nature. For Locke, belief in God is "the foundation of all morality', particularly of the idea of natural equality. God has created us "in [...] a state of equality, wherein all the power and jurisdiction is reciprocal, no one having more than another" (Locke, *On Toleration*, as quoted in Trigg 2007, p. 83). This provided the Christian grounding for the idea of toleration regarding religious matters—something we might take for granted today in Western democracies, but that wasn't obvious at all in Locke's post-Reformation England. Toleration was a matter of natural equality among humans. It was valuable in itself, but also functional to social stability. The only people not deserving to be tolerated in Locke's view are the non-tolerant, because they fail to live up to that standard of human equality. Atheists were a case in point: without religion, they could not have any grounding for the ideal of human equality and therefore for toleration itself. (Roman Catholics were also not to be tolerated, although for more complex reasons. It was, after all, a rather narrow principle of toleration.)

It is hard to deny the importance of (a certain understanding of) equality and toleration in public discourse about bioethical controversies. Seeing one's interlocutors as equal allows everyone to listen to and assess the reasons that others provide, without anyone's voice being granted special or lower status. Here equality refers to the respect or charitable approach owed to all those who join the table with their own values and backgrounds. We are still at the level of procedural values. Robert

Audi and Kevin Smith propose three basic, uncontroversial principles for public discussion and policy making on bioethical issues, which they take both religious (like Audi) and non-religious (like Smith) people would agree on. Two out of these three concern this fundamental ideal of equality as a procedural value. As they stated them, these principles are:

> *The liberty principle*: Government should protect religious liberty to the highest degree possible.
>
> *The equality principle*: Government should accord equal treatment to different religions within its jurisdiction.
>
> *The neutrality principle*: Government should accord equal treatment to the religious and the non-religious" (Audi and Smith 2023).

However, the danger—and indeed the move by Christian thinkers—is to assume that this equality, which so far has been understood as the ground for the procedural value of toleration, also applies to substantive ethical issues, including abortion. For instance, Nigel Biggar says that the Christian religion "offers support for substantive notions of universal human dignity, basic human equality and natural human rights" (Biggar 2015, 231). He appeals to other philosophers and theologians who argue that this view of equality is specific to the Christian tradition. According to Jeremy Waldron, we cannot "shape and defend an adequate conception of basic human equality apart from some religious foundation" (Waldron 2002, 13). And according to Raimond Gaita, "[r]eligious traditions speak of the sacredness of each human being", a concept that does not have a 'secure home' outside of these traditions (Gaita 2000, 5).

This move is implicit in the Christian notion of equality based on a theological view of human nature as God's creation. However, it is not available—at least not without further arguments—outside of it. At this point we are no longer dealing with the procedural aspect of equality that Trigg connected to the Lockean value of toleration. We are dealing with a substantive view about the equal moral status of all human beings. In this Christian or more broadly religious view, this notion of equality necessarily extends to foetuses.

And herein lies the problem with Christian equality. Unlike the procedural principle of toleration, this substantive notion of equality grounding a moral rejection of abortion is not part of many non-religious views on moral status. The convergence between Christian and secular thinkers on the procedural dimension of equality as a ground for toleration does not imply a convergence on the substantive dimension of equality. That is because the understanding of equality that secular thinkers adopt is not based on the Christian idea that every human being (including foetuses) is created in the image of God. To many of those who think abortion is at least sometimes morally permissible, whether that principle of equality extends to foetuses is precisely the issue at stake. To them, human nature itself is not what matters for equality of moral status. Secular thinkers raise and answer a further question that does not arise for Christians, satisfied with universal application of the principle of equality to whoever is created in the image of God. That is: "equality among whom?". On some such secular views, foetuses should not be included. They are not equal in moral status to humans at later developmental stages. For example, many think that foetuses do not have the same moral status as more developed humans because foetuses lack morally relevant features such as self-awareness, a minimum level of rational capacity, and so on. Equality of moral status only applies to entities that have these relevant capacities. Human nature itself, on this view, is not sufficient—and indeed not necessary for having the same moral status as normal adult humans (in principle, entities like sophisticated enough robots or aliens could, on this view, have moral status). This view, of course, comes with its own problems, for instance about the moral status of people with severe cognitive impairments or of newborns, and there is a discussion to be had about these issues. Secular thinkers might or might not have good answers, but they would point out that there is something worth discussing in there.

Now, this is all well-rehearsed in the bioethics literature, and not something I intend to defend or refuse here. Secular thinkers might well be wrong and Christian thinkers might well be right in saying that equality of status, dignity, and therefore rights apply to foetuses and adult humans alike, merely in virtue of being human. The issue is very much up for debate. But to say that the same procedural equality that grounds a principle of toleration is also the substantive equality that grounds the equal moral status of foetuses and fully developed humans is a way of bypassing that debate. It follows automatically for those who share the assumption of a God that created all of us in His image, but not for everyone else.

7.5 Conclusion

Religion has an important role to play in bioethical discussion around abortion because religious values matter to many people. It is important that Christian, Muslim, and any other religious group can participate in public discussions about whether and when abortion is morally permissible according to the respective doctrine. This type of discussion is a fundamental aspect of bioethics, both at the academic level and at the level of societal discussion.

The role of religion in public discussion about the place of abortion in professional settings and about abortion legislations is more problematic. Religion does have arguments, values, and principles that are relevant to the politics and policies on abortion. However, in liberal pluralistic societies, these can contribute to that discussion only once appeals to faith or authority in their support are dropped. If these faith and appeal to authority are distinctive aspects of religion—and I have acknowledged this is a problematic view—then there would be an only marginal place for religion at the table of bioethics in liberal pluralistic societies. Religion can convey important values and norms that exist independently of it, and that as such are open to public scrutiny.

In particular, natural law theory and a principle of equality are the two ways in which a convergence between Christian and secular views on abortion can be created. If that theory and that principle can produce good arguments against abortion, then religion can at best be instrumentally valuable to promote them, but not to ground them. In any case, while these values and norms might well be valid and widely supported, they are also open to rational scrutiny, political negotiation, and, in some cases, defeat. Those supporting the role of religion in bioethical discussion on abortion would need to accept this implication.

References

Audi, Robert, and William R. Smith. 2023. Religious Pluralism and the Ethics of Healthcare. *Bioethics* 37:42–51.

Biggar, Nigel. 2015. Why Religion Deserves a Place in Secular Medicine. *Journal of Medical Ethics* 41:229–233.

Bishop, John and McKaughan, D. 2023. Faith, The Stanford Encyclopedia of Philosophy (Winter 2023 Edition), edited by Edward N. Zalta & Uri Nodelman. https://plato.stanford.edu/archives/win2023/entries/faith/

Camosy, Charles C. 2012. The Role of Normative Traditions in Bioethics. *The American Journal of Bioethics* 12 (12): 13–15. https://doi.org/10.1080/15265161.2012.725349.

Dunstan, G. R. 1988. The Human Embryo in the Western Moral Tradition. In *The Status of the Human Embryo*, edited by G. R. Dunstan and M. J. Sellars, 1–18. London: King Edward's Hospital Fund.

Ford, Norman. 1988. *When Did I Begin? Conception of the Human Individual in History, Philosophy and Science*. Cambridge: Cambridge University Press.

Irving, Diane N. 2000. Abortion: Correct Application of Natural Law Theory. *The Linacre Quarterly* 67 (1): 45–55.

Ling, Trevor. 1980. Karl Marx and Religion. In *Karl Marx and Religion: 10 Essays*, 145–166. London: Palgrave Macmillan. https://doi.org/10.1007/978-1-349-16375-5_10.

Murphy, Timothy F. 2012. In Defense of Irreligious Bioethics. *The American Journal of Bioethics* 12 (12): 3–10. https://doi.org/10.1080/15265161.2012.719262.

Oakley, Justin, C. A. J. Coady, and Lauren Notini. 2023. Religious Pluralism in Healthcare. *Bioethics* 37 (1): 3–4.

Office for National Statistics (ONS). 2021. *Census*. Accessed 22 August 2023. https://www.ons.gov.uk/census.

Quong, Jonathan. 2022. Public Reason. In *The Stanford Encyclopedia of Philosophy (Summer 2022 Edition)*, edited by Edward N. Zalta. https://plato.stanford.edu/archives/sum2022/entries/public-reason.

Rachels, James. 1995. *The Elements of Moral Philosophy*. New York: McGraw-Hill.

Reichlin, Massimo. 1997. The Argument from Potential: A Reappraisal. *Bioethics* 11 (1): 1–23.

Tonti-Filippini, Nicholas. 1989. A Critical Note. *The Linacre Quarterly* 56 (3): 36–50.

Trigg, Roger. 2007. *Religion in Public Life: Must Faith Be Privatized?* Oxford: Oxford University Press.

Waldron, Jeremy. 2002. *God, Locke, and Equality: Christian Foundations in Locke's Political Thought*. Cambridge: Cambridge University Press.

Chapter 8
Abortion, Religion, and Bioethics

Robert P. George and Christopher Tollefsen

Abstract This chapter examines abortion in relation to religion within bioethics and public policy. Drawing on centuries of Catholic philosophical and theological reflection on human nature, dignity, and rights, the analysis avoids appeals to revealed truth or authority beyond reason, extending its relevance beyond specifically Catholic perspectives. The discussion unfolds in three parts. First, it explores the Catholic commitment to natural law and reason, emphasizing the legitimacy of Catholic participation in public debates within liberal democracies on the moral obligations owed to human beings in prenatal development. Second, it argues for the inclusion of voices from faith traditions that may not share Catholicism's robust commitment to reason, asserting their right to participate in public discourse on abortion. Finally, it considers distinctly Catholic claims regarding the morality and politics of abortion, highlighting three claims deemed essential for resolving the issue, which are defensible as truths of natural reason and consistently taught by the Catholic Church.

R. P. George
Department of Politics, McCormick Professor of Jurisprudence, Princeton University, Princeton, NJ, USA
e-mail: rgeorge@princeton.edu

C. Tollefsen (✉)
Professor of Philosophy and Peter McCausland Chair in Civic Engagement, Department of Philosophy, University of South Carolina, Columbia, SC, USA

K. R. Smith (ed.), *Bioethics and Religion*, The International Library of Bioethics 116, https://doi.org/10.1007/978-3-032-33538-8_8

8.1 Reason and Religion

In our own work about abortion and unborn human life more generally, we have relied upon few, and often no, claims that are distinctly religious.[1] Our fundamental view about the morality, including the political morality, of abortion is summarized in four propositions:

- Each of us is, essentially, a human being, a human *organism*. We are not embodied minds, disembodied souls, or some other kind of entity distinct from the living human being whose reflection we see in the mirror;
- Each of us, unless we are the product of monozygotic twinning or human cloning, began our life at conception, understood as the point at which a female oocyte was penetrated by a male sperm giving rise to a genetically and functionally distinct organism—a human being in the embryonic developmental stage.[2] At that point, both gametic cells ceased to exist and were replaced by a new organism with which we are even today biologically continuous and to which we are numerically identical;
- All human beings are equal in fundamental worth and dignity and are, as a matter of justice, entitled to the same protections against killing that we, the authors, and you, the readers, of this chapter are; i.e., it is morally impermissible to kill pre-natal human beings under the same circumstances that it is impermissible to kill human beings who have been born;
- The fundamental purposes for which a state and political authority exist include the protection of all those human beings within its borders from unjust or otherwise impermissible killing and all other unjust forms of violence. Thus, it is not just reasonable, but a political-moral obligation, for the state to protect prenatal human beings from elective abortions.

We have defended each of these propositions at length in various works, and we have defended each on grounds that are entirely non-sectarian and make no appeal to putatively revealed propositions. In doing so, are we acting in some way contrary to, or outside the boundaries of our own religious tradition, namely, Catholicism?

Certainly not. For our own religious tradition makes strong religious claims about the importance of natural reason, both as a truth-attaining faculty, and as providing a common ground to be sought between people of different religious faiths and between religious believers and unbelievers.

[1] See especially Robert P. George and Christopher Tollefsen, *Embryo: A Defense of Human Life* (NY: Doubleday, 2008; 2nd ed. Princeton, NJ: Witherspoon Institute, 2011).

[2] At least one identical twin comes into existence later than conception through some process of either budding or splitting of the early embryo; a cloned human being comes into existence at the moment at which there is a substantial change from a somatic cell of some sort, with its restricted potency for development, to a genuinely totipotent cell, that is, a cell with the primordial capacity to generate all the cells of the early human body, including those of the trophoblast, in the ordered way characteristic of human embryological development.

Both these points are important. As to the first: while the Catholic Church holds that it has been given the charism and authority to teach decisively on matters not only of faith but also of morals, the Church teaches also that the moral truths that she teaches can be known as well by unaided reason. Reason's grasp of such moral truths is a grasp of the "natural law", so-called not because it involves a scrutiny of the natural world but precisely because the use of reason is natural to human beings and is capable of identifying and affirming moral truths.[3]

Thus, for example, while the Church has taught since very early in its history that abortion is to be considered a sin, it has held that this claim is accessible to natural reason and is not matter of revelation only.[4] Moreover, the Church has allowed its own understanding of abortion to be guided by reason. Thus, early treatments of the wrong of abortion in the Catholic Church did not seek to articulate in great detail the difference between early abortion and contraception, although they were recognized to be distinct wrongs. And at various points the Church has similarly distinguished, without denying the wrongfulness of either, between abortion before formation of a human being, and abortion subsequent to such formation. At a time at which biological knowledge was limited or erroneous, this approach was sensible. But after the discovery, by use of the microscope, of essential features of human embryology in the nineteenth century, and especially Karl von Baer's discovery of the mammalian ovum in the 1820s, Pope Pius IX in his 1869 Bull *Apostolicae Sedis moderationi* eliminated any canonical distinction between early and late abortions, implicitly acknowledging the humanity of the unborn human being from conception forward.[5]

This reliance on and acknowledgement of the power of natural human reason has, further, been recognized by the Church as essential for what we might call the possibility of a global ethic, an ethic that transcends religious disagreements, between Catholics and other Christians, Christians and non-Christians, theists and non-theists. The Church recognizes that its authority in matters of morals will not be recognized by those with different or no faith commitments; accordingly the Church *encourages*, for the sake of civil relations between persons in a pluralistic society, the exchange of reasons where crucial matters of morality are at stake. St. Thomas Aquinas put the point succinctly: "...if you are disputing with people who accept no authority, you must resort to natural reasons."[6]

[3]The *Catechism of the Catholic Church*, #1954, quotes Pope Leo XIII, *Libertas praestantissimum*: "The natural law is written and engraved in the soul of each and every man, because it is human reason ordaining him to do good and forbidding him to sin..."

[4]Catholic teaching on abortion dates back to the late first or early second century document the *Didache*, the *Teaching of the Twelve Apostles*, which holds "You shall not slay the child by abortions. You shall not kill what is generated."

[5]For a brief but helpful history of the Church's teaching on abortion, see John M Haas, "The Catholic Position on Abortion," "The Catholic Position on Abortion," *The Linacre Quarterly*. 60: 1993, 67–76.

[6]St Thomas Aquinas, *Quodlibetal Questions* IV q. 9 a. 3c.

So there is an important, even essential, role for reason in bioethical debates, including the debates over abortion, if one is Catholic. Indeed, in a pluralistic society, to abandon reason in favor of exclusively sectarian arguments about abortion is a decidedly un-Catholic thing to do.

But what of those religious persons, Christian or otherwise, whose thinking about abortion is shaped primarily by faith-based considerations? Does the Catholic emphasis on natural reason mean that such religious thinkers should have *no* voice in discussion of the ethics or public policy of abortion? We turn now to this question.

8.2 Religious Reasons in the Public Square

There is a robust tradition in contemporary liberal discourse that holds that only *public reason* is permissible in certain kinds of public debates or as grounds for certain judgments of public policy, of which debate and policy judgments over abortion might well be a paradigmatic instance, touching, as it does, on "constitutional essentials and matters of basic justice." The interpretation of public reason, however, is vexed. For John Rawls and his followers, public reason would seem to exclude even the kind of approach we have discussed in section one of this chapter, on which natural reason may, and indeed must be brought to bear fully on questions related to the political morality of abortion.[7] For on our account, and that of the Catholic tradition more generally, the ethical principles governing abortion are drawn from substantive and rich considerations about basic human goods, integrally conceived, and human flourishing. Ethics, on our view, just is a reasoned account of the principles identifying aspects of human flourishing and further principles that identify acts required by, or inconsistent with, such flourishing, and politics is an extension of ethics that considers the application of such principles to social life.

But for Rawls, political morality, at least in its constitutional essentials, must abstract from substantive conceptions of human good, and must, in order to adhere to canons of public reason, refrain from making use of claims that citizens might reasonably be expected to disagree with. Rawls considers this a duty of civility that should be exercised by citizens in relation to one another.

Rawls's public reason-based restrictions appear to cut against our own natural law based approach to abortion, as suggested both by an infamous footnote in *Political Liberalism*, and a more detailed public reason-based defense of abortion put forward by Judith Jarvis Thomson.[8] We find that part of Rawls unpersuasive and have argued against it elsewhere. We are here concerned with an area where there might seem to be overlap, however, between Rawls and our view, or the Catholic

[7] See John Rawls, *Political Liberalism* (Columbia University Press, 1996), Lecture VI. See also Rawls, "The Idea of Public Reason Revisited," *University of Chicago Law Review* 64 (1997).

[8] See the first edition of Rawls, *Political Liberalism* at n. 32, page 243; see also Judith Jarvis Thomson, "Abortion," *Boston Review*, October 1995. https://www.bostonreview.net/forum/judith-jarvis-thomson-abortion/.

view more broadly. For Rawls's restrictions also cut against the full inclusion of *exclusively* religious voices in public discussion of foundational issues of justice, such as abortion. Religious reasons are, in a pluralistic society, inevitably disputed and apparently sectarian. So Rawls's emphasis on public reason can seem much more plausible in the face of an adherence to a *strictly* religious approach to abortion.

At the same time, there are reasons for hesitation. An important test case for Rawls is the civil rights movement, much of the inspiration, rhetoric, and argument of which was undeniably and very explicitly religious. Thus Rawls suggests, in a slight softening of his position, what he calls a "proviso", that religious voices might permissibly make their case in the language of their substantive faith traditions provided that those views in some way had the backing of public reason as well.[9] A related approach might hold that religious voices must be willing, or must make an attempt, or must be willing to let others make an attempt, at *translation*, at articulating one's religious reasons in terms that might be acceptable to those not of one's faith.[10]

Some critics have argued, convincingly, we think, that such restrictions on religious citizens are likely to be unfair.[11] Although it is true that the most stringent public reason approach restricts the use of substantive non-religious accounts such as our own natural law account, Rawlsian restrictions are more likely to be enforced against religious citizens than against other citizens who hold substantive but not necessarily or specifically religious conceptions of morality.

Now, Rawls's "restrictions" are not, in an important sense, "enforceable": they are duties of civility that citizens have towards one another in a pluralistic society, but are not legally enforced. But perceived failure in regard to duties of civility are surely met with social sanctions, and it is no exaggeration to suggest that citizens who speak in a purely, or even in a heavily, religiously inflected language on matters of fundamental justice are often looked upon and treated with some scorn by non-religious citizens—at least when they are speaking in support of laws or policies, such as pro-life policies, that offend the sensibilities of most secular progressives. And, given the emphasis that we have placed on reason—"public" reason, we think, in the important sense of being available to everyone—it might reasonably be asked of us whether we also think that religious citizens have a duty to refrain from participation in public discussion of disputed questions such as those concerning abortion, at least when using only religious language, and whether social sanctions directed against those who fail in such duties is warranted. For such citizens, the objection might continue, speaking from their sectarian standpoints, cannot be offering to those with whom they disagree reasons that those in disagreement will or can find persuasive or convincing. What then is the point of such religious

[9] Rawls, "The Idea of Public Reason Revisited," 784.

[10] See, e.g., Jurgen Habermas, "Religion in the public sphere," *European Journal of Philosophy*, 12 (2006): 1–25.

[11] See, e.g., Nicholas Wolterstorff, "Why we should reject what liberalism tells us about speaking and acting in public for religious reasons," in P. Weithman (Ed.), *Religion and Contemporary Liberalism* (South Bend: University of Notre Dame Press, 1997), 162–181.

citizens participating in public discussion, except perhaps in order to advance some barely disguised theocratic agenda?

We ourselves are critical of truly theocratic governance, especially (though not exclusively) in religiously pluralistic societies; and we think that there are violations of civility that can emerge from certain forms of religious discourse in the public square. But we note a further problem with such an ethic of religious restraint, namely that it treats some citizens as incapable, from their own conscientiously occupied point of view, of being able to participate in the political project of being governed and *governing*, of participating in the political life of a democratic republic. This is because some citizens, for religious reasons, believe that religion is not merely a, but *the* privileged way of arriving at the truth (perhaps they believe, as do some Christians, that reason has been too corrupted by sin to be trusted). Even to engage in translation, for such citizens, might be in an important sense to fail to be true to their own position, which requires them, as a matter of religious conscience, to articulate their views in entirely religious language using entirely religious reasons.

Perhaps such citizens will be less effective participants in public debate than citizens who engage in "translation", or who embrace a robust natural law understanding of justice and the common good.[12] But can we say that as a matter of political morality they should be thought to be engaged in a violation of civility if they participate in political debate, such as that over abortion and the rights of unborn children, in the only way that they believe is permitted to them in conscience? Or is it rather a violation of civility, and a failure to treat all citizens as equal, to demand restraint of such persons, much less to visit social sanctions upon them when they act conscientiously?

We argue here for the latter position. First, let us grant that Rawls is not incorrect to say that civility requires being willing to give reasons for one's policy proposals, and further requires a commitment to living in a polity in which people are encouraged and allowed to do that. Second, let us likewise grant that for something to be a reason, it must be in some sense publicly accessible; indeed, the idea of something being a "private" reason seems to us oxymoronic. But an important feature of reasons is that the accessibility of reasons requires that some epistemic conditions pertaining to the reason-seekers be met—an epistemic agent must be cognitively mature, sober, adequately educated, etc., in order to recognize some consideration *as* a reason. Sometimes those epistemic conditions can be satisfied by most people; but sometimes they can be satisfied only by people in very specific conditions: the reason for believing there is a starling outside my window might be accessible only, for example, to those who are right now looking out this particular window. Yet the reason for believing there is a starling outside my window is public, nevertheless.[13]

[12] But perhaps not: as Ryan Anderson has pointed out in correspondence, Martin Luther King Jr's use of religious language seems to have been effective with other religious believers who nevertheless initially disagreed with him about civil rights for African American citizens.

[13] Moreover, my testimony that there is a starling outside my window appears to be a reason, and hence a public reason, for others to believe there is a starling outside my window, even if those

Now most, perhaps all, religions (apart from certain strictly gnostic sects) present even their most distinctive doctrinal claims as truths that *would* be recognized by anyone who occupied the appropriate epistemic standpoint. The conditions necessary to occupy that standpoint might be difficult to achieve: they might require being a robust participant in a particular religious tradition, or the presence of infused grace or theological virtues such as faith. But religious claims made by religious citizens in public policy debates are held by religious citizens to be truths that could in principle be recognized by everyone, provided they were in the appropriate epistemic conditions. Thus, insofar as those truths are presented as grounds for some or other choice of public policy, they are indeed presented as public reasons for that policy.

In this, religious claims made by religious citizens are profoundly different from the claims offered by citizens, whether religious or otherwise, that are made without appeal to any reasons. For example, were a citizen to assert "abortion should be legally restricted" (or: "abortion should be legally permitted") and offer no reasons at all for their prescription, this would be a failure to offer public reasons and thus would be a violation of civility. But to our mind, it is no violation of civility, even if it is occasionally or often ineffective, to say "abortion should be legally restricted because the Bible teaches that human life is sacred" (or: "we of the Satanic Temple—a religion recognized by the U.S. Internal Revenue Service—hold that abortion should be legally permitted because one of the seven tenets of our faith is that 'one's body is inviolable, subject to one's will alone'"). Such an approach evinces a willingness to offer reasons, and thus is consistent with a demand for civility in a liberal democratic state, so long as it is accompanied by a willingness to listen to and engage with the reasons offered by others.[14]

But then does anything go? Are there *no* boundaries where religious reasons are concerned? We think there are, but located in a slightly different place than is typically recognized by anti-perfectionist liberal political theorists, such as Professor Rawls. We distinguish, as those thinkers do not, between *purpose* based restraints, and *source* based restraints.

Purpose based restrictions emerge from consideration of the purposes of political authority and its limits. These purposes are primarily concerned with making possible the self-constitution and flourishing of citizens, individually and in community, by providing a stable and just social world within which such flourishing can be pursued. One could thus summarize, from one point of view, the state's purposes, and thus the limits to state authority, as essentially those of justice and peace.

Because one constitutive aspect of human flourishing includes attention to the good of religion, pursuit of the state's purposes includes fostering *opportunities* for religious citizens to purse that good individually and communally; but the purposes

---

others are in no epistemic position to verify the claim for themselves; thanks to Sherif Girgis for pointing this out.

[14] Of course, being reasons does not make them good reasons, and the reasons offered by religious citizens of various kinds can be criticized and rebutted by those who think there are better reasons for some other policy.

of the state do *not* include realizing the goal of *proper religious worship*, the parameters of which states are not as such competent to discern.[15] Thus, a political authority that seeks to impose forms of worship on its citizenry goes beyond the purposes for which political authority exists (purposes which include, we reiterate, appropriate respect for the good of religion). And advocacy of such state imposition is, we believe, uncivil in that it advocates as reasons for political action truths (or, often, falsehoods that are believed to be true) that are inappropriate to the purposes of the state.

By contrast, a citizen of a liberal democracy might well take his or her reasons for advocating some position on justice—say, universal health coverage—or peace—say, a principled opposition to war—*entirely* from his or her religion. Because his or her views, regardless of their source, bear on the proper purposes and uses of political authority and, by our earlier argument, constitute genuine reasons, we think there is no duty of civility for him or her to refrain from making his or her case politically in religious language, nor from further political activity such as: forming alliances with citizens who *do* engage in translation, or who do occupy natural law friendly traditions, in order to advance what he or she believes to be requirements of justice and peace. Thus for all practical purposes, although we think our own natural law based arguments are often likely to be more effective, and more broadly convincing, than purely religious arguments about the status and rights of unborn children (or other contested issues), we do not think a political morality that is sensitive to the requirements of civility must hold that advocates of those purely sectarian arguments must censor themselves or be silenced or shamed by others.

8.3 The Resources of the Catholic Moral Tradition

As we have noted, the Catholic Church places great stock in natural reason, holding that the deliverances of reason are in principle available to all persons. Thus, the Catholic Church holds that there are moral truths available to all, a moral law "written on the heart" of each human person, including those who do not know or accept truths that have been specially revealed. Yet the Church also teaches as a matter of revelation many of these same truths. Why the apparent redundancy?

Thomas Aquinas presented an answer which reflects the Catholic Church's thinking on this matter. It is true, he writes, that much of what the Church teaches as

[15] The Second Vatican Council makes this distinction in *Dignitatis Humanae: On the Right of the Person and of Communities to Social and Civil Freedom in Matters Religious*, #3: "The religious acts whereby men, in private and in public and out of a sense of personal conviction, direct their lives to God transcend by their very nature the order of terrestrial and temporal affairs. Government therefore ought indeed to take account of the religious life of the citizenry and show it favor, since the function of government is to make provision for the common welfare However, it would clearly transgress the limits set to its power, were it to presume to command or inhibit acts that are religious."

morally important for salvation is also knowable through natural reason. Nevertheless, in regard to both theoretical truths, such as the existence of God, or ethical truths, what can be known by natural reason is often grasped, if at all, only with difficulty, or only partially, and is subject to error on the part of us undeniably fallible human beings. Thus, such truths are also taught by the Church in order to preserve her members from such errors as would be detrimental to their salvation.[16]

In this section, we identify three of the moral truths, taught both by the Catholic Church as a matter of revelation and, we believe, knowable by natural reason, that are important resources for thinking about the morality, including the political morality, of abortion. Following our arguments of the previous section, these truths might be deployed in public bioethical debate as matters of natural reason, or as matters of Catholic teaching; civility requires neither purgation nor translation. The truths we address are commonly considered to be part of the Church's teaching on "social justice": truths essential to the common ordering of our lives in society, including political society.[17]

8.4 The Dignity of the Human Person

In Section One, we identified four propositions summarizing our approach to the question of abortion and the rights of unborn children. The third was:

- All human beings are morally equal and are to be afforded the same protections against killing that we, the authors, and you, the readers, of this chapter are; i.e., it is morally impermissible to kill unborn human beings under the same circumstances that it is impermissible to kill born human beings;

This claim is, by our lights, somewhat insufficiently articulated here. For it leaves open the extent and nature of those protections, and their relation to the "moral equality" (that is to say, the equal worth and dignity) of human beings. Much, perhaps all, of the necessary supplementation can be accomplished by introducing the Catholic notion of the dignity of the human person.[18]

According to Catholic teaching, human beings, as rational animals, possess two great, even God-like, powers, the power of reason, and the power of free choice. The possession of such powers, present in the human being by nature and hence from the very moment the human being comes into existence, elevate the nature of the human

---

[16] See Thomas Aquinas, *Summa* Theologiae, I, Q. 1, a. 1. On Aquinas's account, the Church also teaches truths to be believed by faith that go beyond anything available to natural reason, such as the triune nature of God.

[17] Thus, the body of teaching that includes these three principles is commonly referred to as "Catholic Social Teaching."

[18] For a much more extended treatment of the nature of human dignity, and its relation to the norm against killing, than we can offer here, see Patrick Lee and Robert P. George, "The Nature and Basis of Human Dignity," *Ratio Juris* 21 (2008), 173–193.

being far beyond that of the rest of material creation (so far as we are aware). The exercise of those powers might be blocked, in some human beings, by internal or external obstacles; but a necessary condition of being a human being is that one possess these capacities as a matter of primordial radical potentiality. (And so, correcting, say, a genetic or other defect that had caused severe cognitive disability in a human being would not be altering the nature of that individual, it would, rather, be perfecting the individual in respect of his or her nature as a rational creature.)

Thus every human being, equal in nature, is similarly *equal* in possession of human or personal *dignity*, the excellence that comes with the capacities for reason and freedom. And this equally possessed dignity is, in a deep sense, the ground for the norm that no human being may be killed at will and, we think, an even stronger norm: that under no circumstances is an innocent human being's death to be intended (that is, willed as either an end-in-itself or as a means to another end—in other words, made the precise object of someone's choosing).

Catholic teaching thus rejects the common contemporary notion that a human person's life is of only an instrumental value, such that it can be ended when life is characterized by great suffering or believed (by the person whose life it is, or by persons with power over him or her) to be "no longer worth living." This teaching is obviously of profound importance at the end of life, and rules out all forms of "medical assistance in dying." But it is likewise of profound importance at life's beginning: it clearly follows that from the moment at which a new human being exists there is a locus of personal dignity and value whose life may not be deliberately taken for the benefit of another. Catholic teaching militates against any claim that some human being's life, whether because of immaturity, disability, advanced age, or disadvantaged condition, may be held cheaply; all human beings are to be prized in a manner very similar to that argued, on the basis of natural reason, by Immanuel Kant, as possessing dignity, and not price.

8.5 Solidarity

What attitude should be held more broadly with respect to beings possessed of such dignity? Here again, the Catholic tradition of moral reflection has a valuable contribution to make. The Church characterizes the normative relationship that should obtain between persons as one of solidarity: an openness to, and active willing of, the good of other persons.[19] Some clarifying remarks are here in order.

First: to will the good of another is to love the other: love, on this understanding, is more properly volitional than affective—it is an act of will and not mere emotion, though it can, of course, be accompanied by emotion, even strong passion. Second, to will the good of another requires an understanding *of* the good, or, more

[19] For further treatment of the virtue of solidarity, and its application to the context of medicine, see Farr Curlin and Christopher Tollefsen, *The Way of Medicine: Ethics and the Healing Profession* (South Bend, IN: University of Notre Dame Press, 2021), Chapter Three.

accurately, goods, the variegated aspects of human well-being or flourishing that can be pursued for their own sake as constitutive aspects or dimensions of a well-lived life. Such goods include human life and health, knowledge, friendship, and religion, among others. Solidarity thus requires not a thinned out Rawlsian respect for human beings that *abstracts* from the objective good, but a genuine understanding of these constitutive aspects of human flourishing.

Third, what it means to be in solidarity with another depends in important ways on the nature of the community of which the members are a part. Familial solidarity is more intense in the demands it makes of one agent upon another than are more attenuated forms of solidarity such as might exist between members of a business partnership or corporation, a nation, or the community of nations. This insight is a key to understanding an important flaw in Judith Jarvis Thomson's famous "violinist" analogy. It is true that in many cases, the solidarity that is morally required even between strangers would not suffice to proscribe unplugging a violinist who had abducted you to make use of your kidneys. But the relationship of parent, in this case mother, to child, a relationship established by the act of one person bringing into existence another person of equal dignity and worth, is a more demanding relationship than that of strangers.[20] Even if, as Thomson argues, there are abortions that are more like "disconnections" than intentional killings, it is reasonable to think that for any purpose less than preservation of life or defense against significant impairments of health, it is a failure of solidarity to evict one's own child from one's womb at the cost of his or her life. Thus, solidarity, like the dignity of the human person, is a second principle of Catholic Social Teaching that has immediate bearing on the question of abortion, underwriting as it does, "good Samaritan" obligations that are more robust than those imagined by modern thinkers who believe obligations come into existence only when they are voluntarily assumed.

8.6 The Importance of the Common Good

The third principle of Catholic Social teaching we here introduce has already been implicit in our discussion of the purposes for which states and political and legal authority exist. Political society, and legal authority, exist in service of the common good of society. The *political* common good is broadly the set of conditions that are necessary in order to make possible the overall flourishing of persons in that society, both individually and in community. This overall flourishing is, in turn, what we could call the *unrestricted* common good of the society, a common good that

[20] For further discussion of the importance of familial relationships, see Melissa Moschella, *To Whom Do Children Belong? Parental Rights, Civic Education, and Children's Autonomy* (Cambridge: Cambridge University Press, 2016).

transcends the political, or public, common good that consists of those aspects of the common good for which political authority and law exist.[21]

What does Catholic teaching say about the common good, in its restricted or unrestricted senses? In regard to the *unrestricted* sense, the Church teaches that all members of a society, including political and legal authority, must adequately serve the common good in the way(s) appropriate to their role in society. In regard to the *restricted* political common good, the Church teaches that political society must seek the "sum total of conditions by which individuals and groups can more fully and easily achieve their perfection." As we have noted already in this chapter, those conditions include the achievement of both justice and peace within political society.

And the achievement of justice and peace manifestly requires putting into place adequate protections for all innocent human beings who are or might be at risk of violence and harm within the reach ("jurisdiction") of political authority. This principle that the common good must be protected and served is thus at the root of the fourth proposition we asserted at the opening of this chapter:

- The fundamental purposes for which a state and political authority exist include the protection of all those human beings within its borders from unjust or otherwise impermissible killing and other unjust forms of violence. Thus, it is not just reasonable, but a political obligation for the state to protect prenatal human beings from elective abortions.

For this reason we find entirely unconvincing any suggestion that abortion be considered a purely "private" matter. Not just tacitly allowing, but even legally permitting the killing of some class of persons within political society constitutes a radical failure of justice—a failure to treat all human persons within society as equals. Further, it constitutes a radical failure to maintain peace, permitting as it does lethal violence against a disfavored class of human beings. As in its teaching concerning the dignity of the human person and the importance of solidarity between persons, the Church's teaching on the importance of the common good and the moral urgency of its pursuit and promotion seems to us a salutary bulwark against various forms of libertarianism, extreme individualism, and anti-perfectionism that would require political authority to refrain from protecting members of that class.

8.7 Conclusion

To reiterate: all three principles we have identified are, we think, defensible using natural human reason unaided by revelation. Yet clearly a variety of social trends and movements create cultural and intellectual frameworks in which the principles

[21] For discussion of the variegated nature of the concept of the common good in Catholic thought, see V. Bradley Lewis, "Catholic Social Teaching on the Common Good," in Gerard V. Bradley and E. Christian Brugger, eds. *Catholic Social Teaching: A Volume of Scholarly Essays* (Cambridge: Cambridge University Press, 2019), 235–266.

and their implications can become obscured. So the Church takes seriously as necessary to its mission that she continue to teach as divinely revealed these principles of social justice, and to indicate their relevance to contested moral and political matters such as those surrounding abortion. (Notably, few complaints are heard, especially from secular progressives, who object to the Church's teaching on abortion and the dignity of unborn human life, when representatives of the Church invoke the Church's tradition of social justice teaching to advocate for, say, more generous social welfare policies, more aggressive anti-poverty programs, and the abolition of the death penalty.)

We have argued that both when Catholics and others make use of natural reason to argue for these truths, and when Catholics and others invoke the teachings of their faith, they should be permitted, not just legally, but in a way unaccompanied by social sanctions, to participate fully in public discussion of abortion. As we have noted, this does not mean their arguments are immune from scrutiny and criticism; but it does mean they enter public life and policy discussion on terms of equality with those arguing from a purely secular standpoint.

Chapter 9
Secular Ethics Approaches to Medically Assisted Dying

Udo Schuklenk

Abstract Secular ethics approaches to medically assisted dying don't necessarily have to agree substantially on the ethics of assisted dying. What they have in common is a rejection of the authority of other-worldly authorities and their earthly representatives in terms of determining what is morally right and what is morally wrong. Their analyses will be public reason-based as opposed to being based on interpretations of religious scriptures that rely on claims related to these other-worldly authorities. Public reason-based analyses will oftentimes be based on personal autonomy related values systems, or on value systems concerned with an objective list type of human well-being indicators, but they don't have to be limited to those. I discuss significant public reason-based approaches to the ethics of medically assisted dying in this chapter. I also query how, if at all, religious writers engage with public reason-based arguments in the academic discourse on the ethics of assisted dying.

9.1 Introduction

Medically assisted death is here to stay. The number of jurisdictions that decriminalize assisted dying increases steadily, and none of those that have decriminalized this practice has reversed course. Historically senior representatives of monotheistic religions have been opposed to assisted dying on both religious as well as on secular grounds. In jurisdictions where assisted dying has been decriminalized or legalized ethics and policy debates have moved by and large on from the question of whether to permit it at all to the thorny issue of who should be eligible. In most jurisdictions where assisted dying is a legal option, to be eligible for an assisted death one needs to be decisionally capable and—by some definition—terminally ill, but a growing number of jurisdictions, among them the Netherlands, Belgium and Canada, have

U. Schuklenk (✉)
Department of Philosophy, Queen's University, Kingston, ON, Canada
e-mail: udo.schuklenk@pm.me

K. R. Smith (ed.), *Bioethics and Religion*, The International Library of Bioethics 116, https://doi.org/10.1007/978-3-032-33538-8_9

adopted assisted dying regimes that are more permissive. Where to draw the line turns out, unsurprisingly, to be a highly emotive and controversial question.[1]

9.2 Why Religious Authority-Based Ethical Arguments Shouldn't Guide Public Policy in Liberal Democracies

There are sound philosophical as well as historical reasons for why the separation of state and religion is probably a good idea, not least the Thirty Years' War between 1618 and 1648. Ironically, and perhaps surprisingly, part of the reason for why the separation of state and religion is a good idea is that religious freedom in multicultural societies cannot be guaranteed if the state takes a view on the 'true' religion, if any. John Locke in his 1689 *Letter Concerning Toleration* suggested what is today a commonplace view held in liberal democracies, namely that the state's primary function is to concern itself with civil interests, which he defined as 'Life, Liberty, Health, and Indolency of the Body; and the possession of outward things such as Money, Lands, Houses, Furniture, and the like.'[2] (1689–1983). In the context of assisted dying some of these interests make an appearance as a state interest in ensuring the safety and security of its citizens. Liberal democracies' constitutional documents as well as their highest courts have taken their cue from Locke. Influential supreme courts such as the Supreme Court of the United States, or that of Canada, have always been consistently clear that it is not within the courts' remit to act as an arbiter of religion.[3] For instance, Canadian Supreme Court Justice Iacobucci wrote, for the majority opinion, 'in my view, the State is in no position to be, nor should it become, the arbiter of religious dogma.'[4]

In most countries where a medically assisted death is a legal option the health care professions are tasked with the provision of assisted dying services. There are a number of good reasons for this. Arguably the most significant is that the professions are typically very tightly regulated, and professionals are highly trained in the safe provision of medical assistance in dying. This grants an additional layer of security and safety to assisted dying procedures. When we turn to the values of professional associations in the health care context we will notice that, typically, their ethical and professional guidance documents are secular in nature. Back in the 1850s the American Medical Association issued its first code of medical ethics. Even then, the approach to writing medical codes of ethics was decidedly secular. The justification for the code's secular nature was that 'the principles promulgated by this code have been assumed as a common ground upon which every member of

[1] Udo Schuklenk, "Argumenta ad passiones: Canada debates access thresholds to MAiD," *Bioethics* 36, no. 6 (2022): 611–612; Udo Schuklenk, "Rethinking Assisted Dying", *Journal of Social Philosophy and Policy* 41, no. 2, (2024), 327–349.

[2] Locke, John. 1983. *A letter concerning toleration*. Indianapolis: Hackett. (Orig. pub. 1689.)

[3] *United States v. Lee*; *Employment Division, Department of Human Resources of Oregon v. Smith*.

[4] *Amselem*, para. 50.

the Association may stand, without reference to the distinctive principles or doctrines which distinguish the various religious societies existing among the vastly extended and diversified population of our country.'[5]

Neither the state nor professional health care association have any special competencies to adjudicate doctrines of rival religions, and neither do they have the ability to adjudicate disagreements between religious and non-religious views. A corollary of this is an extreme reluctance on the part of liberal democracies to impose religion on their citizens (unlike their government counterparts in theocracies like those in Afghanistan, Iran or the Vatican). While religion isn't forced on citizens and professionals, acting on one's religious convictions also is typically not persecuted as long as the consequences of one's decision are self-regarding. In other words, a religious person's decision not to avail themselves of a medically assisted death that they might be eligible to receive, will be respected. The ethical justification here isn't religion-specific, the rationale is public reason-based, it is fundamentally about respect for autonomous choice and the recognition that we have a right to govern our own affairs in so far as they are self-regarding in their consequences. Hence the long-standing respect decisionally capable Jehovah's Witnesses will find for their decisions to refuse life-preserving blood transfusions when they would be clinically indicated.

Religion based arguments then may have some motivating force for some, or even many, adherents of particular religions, but those same arguments should hold no sway in the development of public policies that affect everyone in society, regardless of their religious convictions, or the absence of such convictions.

Because of these considerations, today most public policy discourses on assisted dying take place on the basis of public reason-based arguments. These are types of reasons that can be embraced by people from many traditions, religious and secular. We will see later how religious authors try—mostly unsuccessfully—to make use of secular terminology in their quest to achieve public policy outcomes that align with their religious views on the matter of assisted dying. But first, what could secular ethical approaches to assisted dying look like?

9.3 Secular Ethics and Assisted Dying

9.3.1 The Ethical Case for the Decriminalisation of Assisted Dying

Broadly speaking there are at least two influential secular approaches to assisted dying. Both result in supporting policy positions to support permissive regulatory assisted dying regimes.

[5] Robert Baker. 2013. *Before bioethics: A history of American medical ethics from the colonial period to the bioethics revolution.* New York: Oxford University Press, p. 181.

The most obvious secular approaches to assisted dying are liberal or libertarian in nature. Both concur that decisionally capable people have a liberty right to end their lives. The argument here relies heavily on an autonomous person-based account, presuming that an assisted dying request was chosen both rationally as well as voluntarily. An autonomous decision seems to require substantial voluntariness, the capacity to reason over one's situation, and the ability to appreciate one's situation in light of one's own values, which entails insights into one's situation and one's options and the consequences of acting on those options. As noted by the authors of a Royal Society of Canada expert panel report on end-of-life decision-making, 'The commitment to autonomy [...], thus quite naturally yields a *prima facie* right to choose the time and conditions of one's death, and thus, as a corollary, to request aid in dying from medical professionals.'[6]

Consequentialists, for another influential approach of moral philosophy, have also good moral reasons for respecting the choices of people who have come to conclude that, intractably so, in balance their experienced quality of life is such that they do not wish to continue living. Quality of life is the key consideration here. Our interpretations of our well-being matter.

Wayne Sumner noted perceptively some time ago that for all practical intents and purposes these two approaches tend to converge when we consider decisionally capable people, because we are the best judges of our quality of life.[7]

The ethically relevant criteria to ground a right to access assisted dying seem to be these: Autonomous choice, decisional capacity, a clear understanding of one's situation, and of one's options, voluntariness, and a request that remains stable over time. Mill had it right in *On Liberty,* we are sovereign when it comes to our bodies. It would not be unreasonable to interfere briefly to assure ourselves that the person wanting to end their life is decisionally capable, and that there is no coercion in play, but otherwise we have no business in preventing their death. This might be the major reason why suicide attempts are not illegal in most liberal democracies. Respecting autonomous, self-regarding choices should be uncontroversial in a good society. Of course, autonomy requires decisional capacity. It matters to ensure that the choice someone makes is truly *their* choice, that it is a choice that is considered and that is their own. Equally important, nobody should be pressured into requesting an assisted death, hence voluntariness is an important consideration. The next condition is one already established in jurisdictions that grant eligibility for MAiD in the absence of terminal illness, namely the requirement that the wish to see one's life ended is stable over time. A rash decision made in a moment of despair should not be acted on, and doesn't deserve societal support. Of course, we need to have a clear understanding of the situation that we are in, and of the realistic alternative courses of action that are available to us.

---

[6] Udo Schuklenk, Johannes J.M. van Delden, Jocelyn Downie, Sheila A.M. McLean, Ross Upshur, Daniel Weinstock, "End-of-Life Decision-Making in Canada: The Report by the Royal Society of Canada Expert Panel on End-of-Life Decision-Making," *Bioethics* 25(s1) (2011): 1–73.

[7] Wayne Sumner. 2011. *Assisted death: A study in ethics and law*. Oxford: Oxford University Press.

It is not a terribly challenging conceptual step from supporting someone's right to a self-directed death to acknowledging also that whoever assists that someone should also be able to do so without legal sanctions. It does not seem plausible to criminalize someone's conduct who volunteers to assist someone else in doing something that is not illegal in their jurisdiction. If I am exercising my legal right to do A, then someone assisting me in doing A should, all other things being equal, also be legal.

9.3.2 The Ethical Case Against the Decriminalization of Assisted Dying

As I indicated in my introductory remarks, secular ethics is not a monolith with *a* firm view on assisted dying, even though the majority of writers on the subject probably take a permissive approach. Having sketched out the affirmative case in the preceding paragraphs, in what follows I shall aim, equally briefly, to sketch out the opposing case. A word of caution, I support decisionally capable people's right to an assisted death, so my skepticism about the opposing views may be noticeable.

Immanuel Kant famously condemned suicide as a denial of one's rational nature. He thought that suicide was a priori not morally choice worthy. Kant's argument for this conclusion is not usually thought among philosophers as meriting too much attention.[8] Kant argued that by ending one's life, one is wrongly giving in to human inclinations (the inclination an individual might feel, for example, to turn their back on a difficult situation they could confront more forthrightly). 'Kant's view is that the person committing suicide takes the easy way out, where reason would dictate that he or she face their problems. It is in this respect that Kant thought that committing suicide turns against rational nature.'[9]

Consequentialists may well find themselves opposing assisted dying, or opposing a particular assisted dying regulatory regime, depending on their empirical assumptions about the potential for abuse, and other potentially or really harmful consequences. Examples of this may include concerns about undue pressure put on people in permissive assisted dying regimes to end their lives for nefarious purposes (eg to access an estate sooner, or to reduce health care costs toward the end of people's lives), worries about slippery slopes, and the impact of permissive policy regimes on vulnerable people.

Then there are a number of approaches that qualify as secular, even though their roots may, in fact, be religious. A case in point is the immensely influential

---

[8] Michael Cholbi, "Kant and the Irrationality of Suicide," *History of Philosophy Quarterly* 17, no. 2 (2000): 159–176.

[9] Udo Schuklenk, et al. (2011) Op.cit.

sanctity-of-life doctrine in medicine.[10] There could be secular sources for the view that human life is so immeasurably valuable that it may never be actively taken, regardless of quality-of-life considerations, and regardless of whether someone decisionally capable makes a considered voluntary request for an assisted death. Historically it appears to be the case that the doctrine has its roots both in religiously influenced medical traditions like the Hippocratic Oath's admonition to the effect to 'neither give a deadly drug to anyone if asked for it', as well as in religious views that suggest that an invisible soul enters us at the moment of conception, which in turn renders us infinitely valuable. It is contested, given that the oldest known version of the Oath itself made reference to various deities, whether it is a religious document, but the admonition to never give a lethal drug may well have had secular roots. Apparently in those days what were doctors were also people occasionally involved in poisoning people for a reward, so the Hippocratic Oath writers aimed to put an end to this. Trust in professionals is a cornerstone of the professional-patient relationship, and this may have been a first attempt at establishing such a trust based relationship, even if the concept of professionalism did not exist at the time. The Oath writers tried to reassure society that doctors are safe, so to speak.

Another arguably secular argument against the moral permissibility of assisted dying is encapsulated in the view that each human life is a gift given by others—sometimes that other can be a deity, but it can also simply mean that it is a gift given by one's biological progenitors. This gift may not be denied by its recipient. An all-too obvious response here could be that anyone is perfectly entitled to reject a gift they were given, particularly if it is a gift they did not even ask for. All of us were conceived by our biological parents, yet none of us were involved in the decision-making process leading up to our conception. It is also arguable at least, as David Benatar has suggested, that all of us would have been better off had we never been.[11]

9.4 Secular Attempts at Rejecting Assisted Dying Popular with Religious Writers

Many religious writers in bioethics—broadly understood as affiliated with religious educational institutions, or self-identified as Christian, Muslim, etc.—have accepted the need to engage in public discourse on the basis of public reason-based arguments, rather than on the basis of their own sectarian convictions. For instance, Lisa Sowle Cahill notes, 'faith language that offers a particular tradition's beliefs about God as the sole warrant for moral conclusions will convince only members of that

[10] Helga Kuhse. 1987. *The Sancity-of-Life Doctrine in Medicine: A Critique*. Oxford: Clarendon Press.

[11] David Benatar, "Why it is better to never come into existence," *American Philosophical Quarterly* 34, no. 3 (1997): 345–355.

tradition."[12] That does not mean, however, that those public reason based, or ostensibly public reason based, arguments are then not deployed to support conclusions that align usually closely with these authors' sectarian convictions. I should note that it is difficult and arguably problematic to speculate about authors' 'true intentions', and so let me say that it is possible that the fact that these religious writers' typically seem to align the conclusions of their public reason-based arguments with the edicts of their religions is merely coincidental as opposed to causal. What matters here, though, is that these types of reasons can be interrogated and criticized by anyone as they do not claim special insights or authority when it comes to supposedly existing deities' views on controversial social matters. There is a level playing field, so to speak, as far as the force or weakness of these arguments and analyses are concerned.

Here is how this plays out in the context of opposition to assisted dying.

9.5 Assisted Dying as a Violation of Human Dignity

It is an interesting phenomenon that the label 'human dignity' has historically been deployed both by assisted dying groups as well as activists opposed to assisted dying in support of their diametrically opposed causes. Remarkably this holds true even in court judgments on the subject matter. A good example of this is a 1993 Supreme Court of Canada judgment. It decided a case involving a patient in advanced stage of amyotrophic lateral sclerosis (or ALS, as it is commonly known). ALS is a fatal degenerative disease. The case itself and the court's decisions is not what is of interest here, but let me note at least that at the time the Supreme Court decided against the decriminalization of assisted dying. What is interesting is that both the majority opinion as well as the minority opinion relied to some extent on 'human dignity' in their respective justifications. The majority wrote, '[a]s members of a society based upon respect for the intrinsic value of human life and on the inherent dignity of every human being, can we incorporate within the Constitution which embodies our most fundamental values a right to terminate one's own life in any circumstances?'[13] In denying the petition the majority opinion thought that 'human dignity' supported the anti-choice case.

The pro-choice minority opinion, surprisingly perhaps, also referenced human dignity, 'the Canadian Charter of Rights and Freedoms has granted the constitutional right to Canadians to life, liberty and the security of the person. It [s.7] is a provision which emphasizes the innate dignity of human existence [emphasis added]...[It] follows that the right to die with dignity should be as well protected as

[12] Lisa Sowle Cahill, "Can theology have a role in 'public' bioethical discourse?" *Hastings Center Report* 20, no. 4(suppl.) (1990): 10–14.

[13] *Rodriguez v British Colombia (Attorney General)*, [1993] 3 S.C.R. 519, [1993] S.C.J. No. 94. at para 129.

any other aspect of the right to life.'[14] One of the judges in the minority opinion asked poignantly, 'What value is there, in life without the choice to do what one wants with one's life?'[15] While it is possible that at least one of the opposing sides is mistaken in their understanding of human dignity, it's also possible that there is something conceptually wrong with 'human dignity' itself that aids its use for diametrically opposing views on policy matters.

Human dignity is quite a popular term in international human rights documents and declarations. Oliver Sensen suggests that is precisely the case because it has no actual meaning, and everyone can project one's own understanding of what human dignity is and what it implies into the term.[16]

Sensen was probably onto something there. A closer look at the literature reveals that 'there is currently no consensus on the moral basis or on the precise meaning of human dignity. It is unclear whether it should best be understood as a basic, or primitive, term of moral language or whether it might reasonably be derived from a moral theory of mainstream appeal.'[17] In the eyes of Kantians, for instance, human dignity is simply reducible to respect for personhood, and, as a corollary, autonomous, self-regarding choice. That seems to map on the quoted view expressed by one of the minority opinion judges. It seems also that this might be what gives rise to pro-assisted dying groups nailing human dignity to their masts. There has been some argument in moral philosophy about the question of whether dignity is a species-related argument or whether it should be limited in its application to persons capable of making rational choices, as Kant suggested. Catholic writers, unsurprisingly perhaps, have a very different take on this. Dignity-related claims for them are decidedly species-related, while Kant may have wanted to limit the use of this concept to people capable of reasoning, ie persons. For Catholic writers, the idea, basically, is that God created the human species as the crown of earthly creation. There is also what is known as the *imago dei* idea, where God somehow self-actualizes through humans. God in a way manifests themselves in humans. The Church itself considers assisted dying an 'offence against the dignity of the human person.'[18] I leave open here whether this is what motivated the judges writing the majority opinion in this court case. Kurt Bayertz suggested many years ago that these kinds of

[14] *Rodriguez v British Colombia (Attorney General)*, [1993] 3 S.C.R. 519, [1993] S.C.J. No. 94. at para 230.

[15] *Rodriguez v British Colombia (Attorney General)*, [1993] 3 S.C.R. 519, [1993] S.C.J. No. 94. at para 211.

[16] Oliver Sensen, "Human Dignity in Historical Perspective: The Contemporary and Traditional Paradigms," *European Journal of Political Theory* 10 (2011): 71–91.

[17] Udo Schuklenk, et al. (2011) Op.cit.

[18] Seper, Franciscus (Cardinal), "Declaratio de euthanasia (Declaration on euthanasia)," *Acta Apostolicae Sedis* 72 (1980): 542–52. (English translation available on the Vatican website, http://www.vatican.va/roman_curia/congregations/cfaith/documents/rc_con_cfaith_doc_19800505_euthanasia_en.html (accessed September 2, 2018).)

appeals to human dignity really are the philosophical equivalent to road stop signs, they aim to pre-empt further debate, in ways similar to the 'Nazi' charge.[19]

An Australian law professor and bioethicist who happens to be a seasoned anti-euthanasia campaigner is Margaret Somerville, of the Australian Catholic University. She has long complained about discrimination by her academic colleagues who tend to brush her off as a religious writer, when in her view her dignity based secular arguments are being unfairly ignored by her secular opponents.[20] However, a closer look at her writings reveals that her understanding of human dignity is nothing much more than a re-labelling of Catholic teachings in secular language. For instance, Somerville writes, 'the concept of dignity must be used to maintain respect for the life of each person, and for human life and for the essence of our humanness, in general.'[21] This does sound a lot like middle-of-the-road Catholic teachings, furnished with the dignity cloak to cover the ideological roots of Somerville's stance.

The function of secular language and seemingly meaningful, deep conceptual terms like 'human dignity' then is merely to camouflage religious ideology. That renders it unsuitable for the justification of normative action guidance in pluralistic societies. It fails the public reason test. What this example highlights is that not every term that has the appearance of a public reason basis actually has such a basis. Somerville uses it to hide a covertly religious conception of dignity. The primary function here is to use it as a cloak to hide other values, values that are much less popular in culturally diverse societal contexts than human dignity is.

As I have argued elsewhere, religious writers opposed to assisted dying end up in a situation where either they follow Somerville's approach and illegitimately employ a covertly religious conception of dignity, or they develop a secular foundation. The limited work that has been done in the direction of the latter suggests that it would not support opposition to assisted dying.[22]

Given the foregoing contributors to normative debates on assisted dying would do well to avoid talk of human dignity. It probably is not entirely coincidental that human dignity today is used by all and sundry in support of their political causes, precisely because it essentially lacks content, while projecting depth and rich meaning.

---

[19] Bayertz, Kurt. 1995. "Introduction: Sanctity of Life and Human Dignity", in Bayertz, Kurt (ed.). *Sanctity of Life and Human Dignity*. Dordrecht: Kluwer, ix.

[20] Somerville, Margaret, "Judge me by my ideas, not my religion," *Globe and Mail* November, 242,013. https://www.theglobeandmail.com/opinion/judge-me-by-my-ideas-not-my-religion/article15577208/ (accessed February 6, 2019).

[21] Somerville, Margaret, "Defining human dignity," *The Gazette* (Montreal), November 22, 2009. https://www.catholiceducation.org/en/controversy/euthanasia-and-assisted-suicide/defining-human-dignity.html (accessed September 5, 2018).

[22] Russell Blackford, Udo Schuklenk, "Religion at Work in Bioethics and Biopolicy: Christian Bioethicists, Secular Language, Suspicious Orthodoxy," *Journal of Medicine and Philosophy: A Forum for Bioethics and Philosophy of Medicine* 46, no. 2 (2021): 169–187.

9.6 The Vulnerability Trope

'Vulnerability' has become another popular rhetorical tool in the anti-choice agenda of religious writers. Much like human dignity it is something that has a secular, public reason-based ring to it, and much like human dignity it is something with strong emotional appeal. After all, we all probably know fellow humans (possibly some non-humans, too) whom we consider vulnerable and in need of our special protections. There is a fair amount of literature both in moral philosophy as well as in bioethics on who exactly is vulnerable, and what it is that makes them vulnerable. What makes 'vulnerability' probably appealing to religious bioethicists is that it is sufficiently vague to permit a writer to project their needs into it.

Margaret Somerville, for instance, writes, 'And society's agreement to help elderly and vulnerable persons to kill themselves or to allow physicians to kill them sends a powerful message that their lives are not worth living.'[23] Of course, societal respect for such individual choices takes no stance at all on whether someone's life is not worth living. Rather, it expresses support for someone who has made that determination and wishes to see their life ended. Respect here is paid to the decider and to their personhood and autonomy. Why such a person's perceived or real vulnerability should lead to a withdrawal of that respect is unclear. In a similar vein Joel Hodge, a theologian at the Australian Catholic University, goes, like Somerville, also after decisionally-capable elderly people and hopes to see their agency removed to make choices he disapproves of. He wrote, 'It is easy to underestimate the vulnerability of the elderly and dying, and the delicate human dynamics involved in caring for them.'[24] Another Catholic writer, Charles Camosy, has identified a different group of decisionally-capable people[25] whose agency he'd like to set aside in order to protect them against their own considered choices, namely 'vulnerable, depressed people.'[26] Like Somerville he aims for broad-sweep inclusivity. This should not come as a surprise, in an ideal Catholic world nobody would be able to access an assisted death. The vaguer the terminology deployed, the more people it might be able to encompass. Vulnerability assessments here do not take place based on an individual's situation, but on their membership in a group. Oddly, numbers

---

[23] Somerville, Margaret. 2014. Read and vote: Should patients be allowed to request suicide? *Globe and Mail*, https://www.theglobeandmail.com/opinion/read-and-vote-should-patients-be-allowed-to-request-suicide/article21106527/ (accessed September 2, 2018).

[24] Hodge, Joel. 2017. Assisted dying laws pose a grave risk for the vulnerable. *Sydney Morning Herald*, https://www.smh.com.au/opinion/assisted-dying-laws-pose-a-grave-risk-to-the-vulnerable-20170904-gyaim0.html (accessed September 2, 2018).

[25] Rooney, William, Schuklenk, Udo, van de Vathorst, Suzanne, "Are concerns about irremediableness, vulnerability, or competence sufficient to justify excluding all psychiatric patients from medical aid in dying?", Health Care Analysis 26, no. 4 (2018): 326–43.

[26] Camosy, Charles. 2014. Right to die, or duty to die? The slippery-slope argument against euthanasia revisited. *ABC Religion and Ethics*, http://www.abc.net.au/religion/articles/2014/09/01/4078456.htm (accessed September 2, 2018).

suddenly matter to people who otherwise are not known for their consequentialist leanings. The more people who are denied their agency, the better.

What is it that all of these writers have in common? None of them bothered defining 'vulnerable person' or 'vulnerability'. One would think this may have been important, given that 'vulnerability' serves as the crutch here that is used to deny agency to decisionally-capable people. Their arguments trade entirely on the vagueness of the term and on their readers' empathy toward vulnerable people. Rosamond Rhodes, a secular ethicist in the United States, has been highly critical of these kinds of rhetorical strategies. She rightly sees unjustifiable strong paternalism at work. Rhodes wrote, 'Instead of trying to respect the autonomy of others by presuming that they are autonomous and trying to see their choices as reasonable from their perspective, classifying people as "vulnerable" denies them respect.'[27]

Rhodes is not alone in her concerns about the use of terms like 'vulnerable'. In a recent revised edition of its influential research ethics guidelines the Council for International Organisations of Medical Sciences (probably better known to many under its acronym CIOMS) explains why it has chosen to eliminate the category 'vulnerable group' altogether, '"A consensus emerged in recent publications that vulnerability can no longer be applied to entire groups."[28]

To avoid misunderstandings of this analysis. I am not suggesting that it is impossible to define 'vulnerability' meaningfully. I am also not suggesting that individual people cannot be vulnerable to such an extent that their capacity for agency is questionable. What I am suggesting is that the label 'vulnerable' has been hijacked in the described instances by religious writers on the lookout for secular terminology in order to deploy it in support of orthodox theological policy stances.

9.7 Slippery Slope Arguments

Unlike 'vulnerability' and 'dignity' slippery slope arguments are more technical in nature, they are more clearly defined, and because of that feature, it is easier to address them in public reason-based discourses. In the context of medical assistance in dying, unsurprisingly perhaps, they are always deployed to warn about real or perceived dangers of the decriminalization or legalisation of assisted dying, or the inclusion of particular categories of patients.

I am referencing again some of the Catholic authors we have come across already. Lest this is held against me, I should note that these are high-profile contributors to public debate in their respective countries (and other countries). An example as

[27] Rhodes, Rosamond, "Rethinking research ethics," *American Journal of Bioethics* 5, no. 1 (2005): 7–28.

[28] CIOMS. 2016. International ethical guidelines for health-related research involving humans. Geneva: Council for International Organizations of Medical Sciences. https://cioms.ch/wp-content/uploads/2017/01/WEB-CIOMS-EthicalGuidelines.pdf (accessed September 5, 2018).

good as any is a semi-popular article by Margaret Somerville.[29] Somewhat more sophisticated is an article by Catholic theologian Charles Camosy.[30] He claims that, "when euthanasia is legalized in cultures where the values of autonomy and consumerism hold sway, we soon end up with the kinds of deaths that almost no one wants." This is a standard slippery slope argument. Euthanasia is legalized in an autonomy affirming society—i.e. most every liberal democracy—and soon we find ourselves on a slippery slope toward a situation where we end up with deaths almost no one wants. Everything Camosy says here is baseless, unless with 'that almost no one wants', he suggests that almost everyone is a Catholic who shares the views of his Church on this subject matter. But yes, for the purpose of agitation on a mass media platform such kinds of arguments have a strong emotional appeal, especially if a bit of vulnerability and dignity is thrown into the mix for good measure.

On this occasion, both Somerville's contribution as well as Camosy's analysis are presented on an Australian broadcaster's website. Their outputs are squarely aimed at a mass audience. I should note, though, that Somerville has advanced similar arguments also in her academic contributions.[31]

Examples of slippery slope arguments affect many discussions of assisted dying regulatory regimes. A case in point, Canada. The country introduced legislation that made medically assisted dying available in 2015. Writers opposed to assisted dying insist that Canada has since then well gone down the slippery slope from a restrictive regulatory initial regime when it legalised euthanasia to considerations, at the time of writing, to opening assisted dying up to decisionally capable people with mental health issues, and people with disabilities. However, the history of the evolution of Canada's assisted dying regulatory regime matters. In its decision declaring the criminalization of assisted dying unconstitutional, the Supreme Court of Canada outlined permissive criteria for access to assisted dying.[32] The country's politicians ignored those criteria when they introduced Canada's first assisted dying law. Unsurprisingly, patients sued in the courts in order to be able to exercise their constitutional rights. They won their legal battles each time they went to court. Eventually Canada's government conceded defeat and promised to align its revised legislation with the Supreme Court criteria. If one did not know this background one could be forgiven for wondering why Canada moved so quickly to change access thresholds to MAiD to include more categories of patients. It has also been noted that since the introduction of assisted dying the number of people dying an assisted

[29] Margaret Somerville, "The importance of stories in the euthanasia debate," 2017, February 28 *ABC Religion and Ethics*, http://www.abc.net.au/religion/articles/2017/02/28/4627671.htm (accessed September 2, 2018).

[30] Charles Camosy, "Right to die, or duty to die? The slippery-slope argument against euthanasia revisited," 2014, September 01. *ABC Religion and Ethics*, http://www.abc.net.au/religion/articles/2014/09/01/4078456.htm (accessed September 2, 2018).

[31] Margaret Somerville. 2014. *Death talk: The case against euthanasia and physician-assisted suicide.* Second edition. Montreal/Kingston: McGill-Queen's University Press: eg p. 31, p. 182, p. 194.

[32] Jocelyn Downie, Udo Schuklenk, "Social determinants of health and slippery slopes in assisted dying debates: lessons from Canada," *Journal of medical ethics 47* (2021): 662–669.

death has increased each year. While this is correct, on its own that doesn't demonstrate that there is a slippery slope at work. One has to be opposed to assisted dying in the first place in order to find a slippery slope type fault with a service that an increasing number of people are taking up.

Camosy found himself an expert on a Fox News program in the United States where the false claim was peddled that Canada would introduce an assisted dying regime that would permit minors to be euthanized. That may or may not have constituted a slippery slope, if it had been true, but this actually was not considered in 2022 when the program was broadcast, and it has not been considered since.[33]

9.8 Conclusions

Secular ethics approaches to assisted dying can—in principle—support both the legalization of assisted dying as well as being opposed to this practice. Their arguments and analyses are public reason based and so can be debated publicly in multicultural liberal democracies by people who hold multitudes of normative convictions, secular and religious.

Christian writers' contributions in these debates have acknowledged that the views they aim to defend need to be public reason derived, or at least they need to appear as if they were public reason derived. A closer look at some of the conceptual tools in their philosophical armament suggests a penchant for the use of vague but pleasant and conceptually deep sounding terminology that hides the actual religious roots of their policy recommendations. While this is a problematic rhetorical strategy, an obvious advantage of such an approach is that it is suitable for a public reason-based evaluation and critique. At the same time, in the square of public opinion such a rhetorical campaign strategy may be successful, regardless of what academics make of labels like 'dignity' and 'vulnerability'.

[33] Euthanasia program is what happens when autonomy goes nuts: Charles Camosy | Fox News Video.

Chapter 10
Killing Off the Elderly, Sick, and Other Expensive Patients: Religious Voices in Public Bioethics Debate

Mark J. Cherry

Abstract This chapter critically examines the ethical implications of euthanasia, physician-assisted suicide, and Medical Assistance in Dying (MAiD) within both secular and Christian frameworks. It addresses the growing acceptance of these practices in Western bioethics, particularly in jurisdictions where the scope of eligibility has expanded beyond terminally ill patients to include those experiencing the challenges of old age, personal suffering, or financial hardship. From a Christian perspective, these practices are unequivocally condemned as self-murder and an affront to God's dominion over life and death. By integrating both philosophical and theological insights, the chapter argues that significant spiritual and moral harms arise from physician involvement in such practices. It also explores the transformation of medical ethics and the professional identity of physicians as they navigate the permissibility of euthanasia and assisted suicide, warning of long-term negative consequences of embracing such practices within the medical profession. The chapter concludes by advocating for a more robust dialogue between secular and religious moral perspectives, arguing trenchantly for the inclusion of religious viewpoints in bioethical discourse, particularly in shaping healthcare policies related to end-of-life care.

10.1 Introduction

Is it wrong for physicians intentionally to kill patients if those individuals consent to be killed? What about physicians helping patients to kill themselves? In the Western world, the answer to such questions seems to be increasingly negative. Indeed, for much of secular bioethics the preferred language to describe such practices is something like "Medical Assistance in Dying" (MAiD), because active euthanasia and physician assisted suicide no longer raise important secular ethical

M. J. Cherry (✉)
St. Edward's University, Department of Philosophy, Austin, TX, USA
e-mail: markc@stedwards.edu

K. R. Smith (ed.), *Bioethics and Religion*, The International Library of Bioethics 116, https://doi.org/10.1007/978-3-032-33538-8_10

concerns, beyond ensuring that the patient has in some sense voluntarily chosen this method of dying or that the patient's physical and mental condition is beyond being able to experience the goods of this world with surrogate decision makers affirming euthanasia on his behalf.

While such practices are often rhetorically presented as a last option for terminally ill patients experiencing irremediable physical pain, such is not necessarily the case. Consider, for example:

- May 10, 2018: David Goodall, an Australian scientist ended his life in Switzerland. He was 104 years of age but was not reported to have had a terminal illness nor did he claim to be suffering or dealing with significant pain. According to the *BBC News*, at his request, the Life Cycle clinic in Basel gave him an infusion of Nembutal, a barbiturate designed to end his life ("David Goodall: Scientist, 104, ends his life in Switzerland" 2018).
- December 2022: Les Landry, a 65-year-old Canadian requested MAiD. According to media reports, Mr. Landry used a wheelchair, had epilepsy and diabetes, and acknowledged that the reason for applying for MAiD was financial. Between a recent reduction in his income and an increase in his rent there was a possibility that he would become homeless (Pennock 2022).
- In an interview, several months before utilizing MAiD, Jennyfer Hatch, a 37-year-old Canadian woman, reportedly gave an interview in which she indicated that she was unable to obtain adequate care for Ehler-Danlos Syndrome. She is reported to have concluded that it would be "far easier to let go than keep fighting" (Pennock 2022).

In Canada, eligible patients are not required to be terminally ill, and rather than significant physical pain the most cited reason for active euthanasia and physician assisted suicide is the loss of the ability to engage in what the individual found to be meaningful activities (Health Canada 2023, p. 6).

The framing question of this volume regards whether religious voices should play any role in bioethics and health care policy debates. As I argue, such a question regards foundational concerns in moral epistemology. As a libertarian (see Cherry 2005/2015), I have argued that secular philosophical moral analysis cannot demonstrate why laws favoring robust prohibitions, such as regulations against euthanasia and assisted suicide, are justified; secular moral analysis similarly provides no justification whatsoever for taxpayer financing or laws requiring physician, other health care worker, or hospital participation in such practices. That argument is summarized in Section II. Through the Grace of God, I am also an Antiochian Orthodox Christian. I know euthanasia and physician-assisted suicide to be deeply sinful practices that, save for God's Grace, place one at real eternal risk. Physician-assisted suicide, euthanasia, and other forms of medical termination are directed against basic human goods and deny God's dominion over the world. Such a death sunders one from the Christian duty for humility and submission to God even unto death. In such matters, the terminology of "murder" and "self-murder" are clear, to the point, and help capture the depth of the spiritual matters at stake. These and related religious and secular implications of killing patients are explored in Section

III. Yet, as I argue throughout the chapter, the deep wrong associated with euthanasia and assisted suicide cannot be fully understood outside of knowledge of God and His commands. Epistemology matters.

10.2 Secular Ethics (and Secular Bioethics): Morality as a Do-It-Yourself Project

> We shall be free; th' Almighty hath not built
> Here for his envy, will not drive us hence:
> Here we may reign secure, and in my choyce
> To reign is worth ambition though in Hell:
> Better to reign in Hell, then serve in Heav'n.
> (Satan, John Milton, *Paradise Lost*, Book 1, lines 259–263).

> So farewel Hope, and with Hope farewel Fear,
> Farewel Remorse: all Good to me is lost;
> Evil be thou my Good...
> (Satan, John Milton, *Paradise Lost*, Book 4, lines 108–110).

Christianity's opposition to euthanasia and assisted suicide is grounded in its knowledge that such actions violate the prohibition against murder (including self-murder). That an individual consents or otherwise signals willingness to die, as in cases of voluntary active euthanasia, does not change the deep sinfulness of such practices. Killing persons radically harms one's relationship with God. Killing another to avoid the burdens of caring for that individual while he or she is dying or to help that person (or oneself) escape the indignities, sufferings, or other challenges at the end of life would be seriously spiritually harmful. As H. Tristram Engelhardt, Jr. explores the point:

> The exemplars of good living and dying, the saints, live and die in a way that has no place for physician-assisted suicide or euthanasia. They have never acted to avoid suffering by suicide or euthanasia. One learns from them, from their lives and deaths, how to draw the lines between murder, suicide, and humble self-sacrifice. What is offered is an experiential knowledge of how to aim at union with God. This experiential, noetic knowledge of progress from oneself to God opens the door to understanding why one should voluntarily accept suffering. (Engelhardt 2000, p. 329)

Only God is the Master of life and death. Facilitating one's own death to avoid being a burden to oneself or others has always been recognized as a rejection of Christian asceticism, of the ways in which we must emulate Christ (Engelhardt 2000, p. 330). Christians must take on Christ, including His submission unto even suffering and death (Romans 6). Consequently, Christianity has always recognized suicide and euthanasia as serious violations of the commandment against killing. Such practices reject the humility and holiness of Christ's life and death (Engelhardt 2000, p. 331). What happens to such moral and spiritual knowledge once God is removed from the equation?

As Elizabeth Anscomb (1919–2001) and John Milton (1608–1674) recognized, albeit in rather different ways, absent a canonical point of view—that is, a binding and fully objective standpoint that is not conditioned by a particular society, culture, or place in history, such as God would provide—moral claims cannot be fully justified. Without a canonical foundation, moral considerations—no matter how deeply they are felt—reflect no more than socially, culturally, and historically conditioned judgments. Without a unique and binding normative perspective, ethical concerns shatter into numerous incommensurable perspectives with no in principle means to choose between them. Moral pluralism is the nature of the human condition.

As Anscomb appreciated, once moral claims are made without knowledge of or reference to God and His commands, it is doubtful that they will be even remotely compelling. It is "as if the notion 'criminal' were to remain when criminal law and criminal courts had been abolished and forgotten" (1958, p. 6). Without God, concepts of "immoral action" are analogous to concepts of "illegal action" in the absence of legislators or police to identify illegal acts, much less courts or prisons to adjudicate and punish individuals for them (1958, p. 6). Without God as the author and justification of morality, as well as the source of ultimate sanctions (e.g., rewards and punishments), the significance of supposedly immoral actions is only the significance that specific persons, cultures, or societies attribute to such actions. Acting immorally in a secular sense possesses little significance beyond prudential concerns, such as being caught and punished.

As an example, consider the way in which traditional Christian appreciation regarding the immorality of sexual activity outside of the monogamous marriage of man and woman have ceased to be plausible for large segments of Western society. Having rejected the Christian God and knowledge of His commands, Christian concerns regarding sexual morality have been rejected as well. Casual consensual recreational sexual activity is regarded as at least morally permissible, perhaps virtuous, and good. The prudential implications of this widespread moral judgment are significant. Among the most obvious potentially negative consequences of sexual activity outside of traditional forms of marriage are the significant financial and medical costs of sexually transmitted disease, the distinct possibility that one will pass on such diseases to others, and an increase in the overall percentage of children born outside of wedlock. Individuals routinely pass on the costs of their sexually transmitted diseases to innocent third parties through increased insurance premiums and taxpayer financed healthcare, and there are well-documented disadvantages for children (social, emotional, psychological, and financial) raised outside of marriage (Cherry 2016). Once such outcomes are socially accepted, or ignored, there are no longer any significant grounds for secular morality plausibly to conclude that sex outside of traditional forms of heterosexual marriage is morally wrong. Provided that it is consensual, sexual activity provides just another itch to scratch. Each person defines his own understanding of the goods of such practices for himself with consenting others.

Secular morality cannot even provide any definitive judgment against what were once commonly held to be immoral sexual practices, such as adultery. At best, secular morality experiences adultery as the breaking of a promise to be faithful to one's spouse, presuming that one has made such a promise. (Some couples do not make such promises, negotiating the conditions for "open marriages", with extra-marital sexual liaisons expected and accepted.) Here, the meaning of adultery as a wrong-doing cannot be articulated except in terms of the actual impact of the broken promise on the persons intimately involved. Even then, the supposed wrong is conditional and focused on prudential interests: if you want to have a good marriage and if your spouse does not approve of adultery, then it would be prudent to avoid committing adultery, or at least to avoid being caught. If one recognizes faithfulness to one's spouse as morally correct behavior, and if such actions accord with one's prudential choices, then one probably ought to so act, otherwise it is unclear why one would bother.[1] Again, the morality of such practices is created by those involved.

Without God, or at least a definitive and fully binding God's eye point of view, morality has no specific content, there are no ultimate sanctions for immorality, and thus no final motivation to be moral. While the morality of sexual activity outside of the monogamous marriage of husband and wife provides useful illustration, the conclusion is similar for moral judgments regarding euthanasia and assisted suicide. Having rejected knowledge of God and His commands, advocates no longer appreciate traditional restrictions against the killing of other persons and self-killing as morally relevant, provided that such individuals consent, personally or through surrogates, to be killed.

Rebranding the killing (or assisted self-killing) of patients as medical care involves a radical shift in the foundational moral understandings traditionally embodied in medicine (see, e.g., Pellegrino 2001). Such practices require us to rethink of the medical profession without its traditional moral, professional, or even legal restraints against intentionally killing patients. As part of a 1999 philosophical symposium on physician assisted suicide, Paul Weithman (1999, p. 569) noted that:

> Serious philosophical and sociological reflection is needed to determine what virtues physicians must have if physician-assisted suicide is to be perceived as a legitimate form of medical care for the dying. It is surely an open question in moral psychology whether those virtues can be sustained and transmitted to young doctors by a medical profession and a legal system no longer committed to the traditional prohibition on action with the intention of ending the lives of their patients.

As physicians are taught that euthanasia and assisted suicide fall legitimately among the usual menu of treatments that they may (or are required) to offer patients and from which patients may choose or demand from doctors, the professional ethos of medicine will dramatically change.

[1] Arguing for a similar point, Jürgen Habermas recognized that the normatively charged concepts of Western moral philosophical analysis are routinely appropriated from Christianity but without God to secure their meaning, force, or truth. As he notes: "It also promoted an assimilation of genuine Christian elements by philosophy. This process of assimilation led to such normatively charged networks of concepts as responsibility, autonomy and justification, history and remembrance, rebirth, innovation and return, emancipation and fulfillment, renunciation, internalization and incarnation, individuality and community" (Habermas 2008, p. 110).

Killing such patients, for example, comes to be thought of as morally no different than putting down a dog whose life has become, in our judgment, too painful. As Gerald Kelly (1957, 118) summarizes the point:

> It likens man to a mere animal; it makes pain the greatest evil in the world. It ignores the fact that no one suffers save through the will of God; that through suffering a man can beautify his character, atone for his sins, take a special part in the sublime work of the Redemption, and win for himself an eternity of glory. This does not mean, of course, that Christians deify pain and sit idly by while men suffer. There is also something Christlike in alleviating pain, in helping the sufferer, and in trying to conquer disease. But there is nothing either Christian or Christlike in killing the sufferer to relieve his suffering.

Judgments that some human lives are less worthwhile due to age, illness, or pain, shifts medicine's focus and application. Yet, without God secular morality cannot provide any definitive moral judgment against such killings; indeed, many have come to find the euthanizing of patients both personally virtuous and professionally fulfilling (see, e.g., Frolic and Oliphant 2022; Downie 2022; Frolic and Miller 2022).

At stake is a foundational epistemological challenge. There are deep and irresolvable tensions among various possible, competing, and incommensurable moral perspectives. Significant moral pluralism exists—pluralism in moral theory as well as in commitment, intuition, and judgment. Secular reason is unable to set aside such pluralism. Instead, it always confronts an irreducible plurality of competing moralities, principles, or perspectives, as well as divergent accounts of moral rationality grounded in various rankings and definitions of cardinal values, right-making conditions, assertions of virtue or vice, and competing accounts of human flourishing with no way of choosing among such moralities that does not already assume what must be proved. Divergent rules of moral evidence and valid ethical inference in turn shape competing theoretical accounts and practical judgments. Not even an appeal to the consequences of actions will resolve such controversies: different understandings of harms and benefits ground diverse understandings of undesirable or desirable consequences. In general secular terms, it is impossible to break through these permanent debates to moral truth without simply begging the question (see Engelhardt 1996, Chaps. 1–3). In short, without access to a canonical, content full, morality, such as knowledge of God would provide, to resolve such debates, there is no definitive secular moral guidance to help us choose among competing accounts of the right, good, virtuous, or just.

So, should euthanasia be understood as a harm or a benefit, a virtue or a vice, just or unjust? The answer depends on one's background moral perspective and interests. Secular epistemology cannot ground a definitive answer. Without God, or at least a uniquely true and binding God's eye perspective, to secure moral content, moral knowledge claims are deflated. As noted, they represent no more than socially and historically conditioned judgments, culturally relative expressions of the good or virtuous, assertions of political ideological commitments, or personal intuitions regarding right and wrong. Secular moral philosophy cannot ground knowledge of bioethical absolutes. Without God, morality has become a do-it-yourself project. Each person, much like Milton's Satan in *Paradise Lost*, defines ethical standards for himself.

Given such irremediable pluralism, general secular reason cannot even demonstrate a universal, content-full, binding morality to secure the justification for political authority to impose a particular moral vision. This is why, rather than providing an authoritative bioethics to guide medical practice and health care policy, secular bioethics all too often regards ideological rationalizations for specific forms of political advocacy: court orders or legislative lobbying for laws designed to establish a particular moral point of view at law (Cherry 2024). In terms of general secular reason, legitimate political authority cannot be drawn from assertions of consensus, ideal theories of rational action, deep moral intuitions regarding human rights, or cardinal moral concerns, such as equality and fairness, much less from the supposed goods of democracy or democratic deliberation.

Insofar as one wishes to avoid simply imposing a specific moral perspective through force, the resolution of moral controversies turns not on which group manages to establish its favored moralistic legislation, but rather on whether parties to particular medical interactions have agreed to participate. Given such moral diversity, the free market and the free association of persons, each seeking to secure his own personal and social ends, is central for understanding morally authoritative human choice. Market transactions and contractual relationships draw moral authority from the agreement of the participants to be bound by their choices. The parties to the transactions themselves convey authority to the enforcement of the specified conditions. Here, permission becomes the ground of the moral justification of such collaboration (Engelhardt 1996, Chap. 3). Morally legitimate political authority is derivative of the actual agreement of actual persons. It is created through the collaborative agreement of the parties themselves. The actual agreement of actual persons limits the moral authority of others, including the state, to interfere with the free interaction of consenting persons, including consensual actions that others find morally reprehensible.

Consequently, laws banning voluntary euthanasia and consensual assistance with suicide are without definitive secular moral ground; similarly, laws requiring physicians, nurses and other health care workers, or hospitals to participate in such practices, or taxpayers to finance them, are unjustified acts of coercive force.

10.3 Killing Patients and Helping Patients Kill Themselves: Spiritual and Practical Realities

The spiritual harm to physicians who euthanize patients or who aid and abet patients in killing themselves is significant. Physician-assisted suicide and euthanasia are not neutral choices. That the patient desired or requested death as part of his "medical" care does not change the spiritual significance of the act. Christianity has always appreciated just how significantly murder and self-murder can sunder one's heart from its proper focus on God. As Engelhardt notes: "The traditional Christian life has always experienced such a death as a separation from the humility and holiness of the life and death of Christ. This opposition to suicide, assisted suicide, and euthanasia is rooted in the experience of the Christian life as a life directed to humility" (2000, p. 331). Assisted suicide and euthanasia involves a purposeful turning away from God and the goodness of His gifts. "Surely you know that you are God's temple, where the Spirit of God dwells. Anyone who destroys God's temple will himself be destroyed by God, because the temple of God is holy; and you are that temple" (1 Corinthians 3:17). Death should be approached with humility and repentance, with prayer and confession. Willful killing of suffering or dying patients, of oneself or others, is rejected.

Borne with humility and submission, however, even suffering can aid in our struggle to craft a life turned wholeheartedly towards God. As St. John Chrysostom notes, "So that I did not say rashly, that these tribulations themselves mark out to us hopes of a resurrection, and make those who are tried the better; for he saith, 'as gold is tried in a furnace, so an acceptable man in the furnace of humiliation' (Ecclus. II.3)" (1994, p. 340). Or, as Mark the Ascetic (sixth century) argued, "The mercy of God is hidden in sufferings not of our choice; and if we accept such sufferings patiently, they bring us to repentance and deliver us from everlasting punishment" (1979, p. 136). Do not misunderstand. Suffering is not intrinsically valuable; nor is it to be sought for its own sake. However, approached appropriately, with proper spiritual guidance, it may play an instrumental role in developing one's spiritual self-discipline. Responding rightly to sickness and suffering, even to death, can help one to accept God's will and to turn more fully to Him (Engelhardt 2014, p. 253).

While the commandment "Thou shalt not kill" (Exodus 20; Deuteronomy 5) straightforwardly rules out euthanasia, the early canons of the Christian Church also made clear that suicide was self-murder and, thus, similarly forbidden. Suicide even raised questions regarding the permissibility of providing the usual funeral offerings on behalf of the deceased. Question XIV from the 18 Canons of Timothy states:

> If anyone having no control of himself lays violent hands on himself or hurls himself to destruction, whether an offering ought to be made for him, or not? Answer: The Clergyman ought to discern in his behalf whether he was actually and truly out of his mind when he did it. For oftentimes those who are interested in the victim and want to have him accorded an offering and a prayer in his behalf will deliberately lie and assert that he had no control of himself. Sometimes, however, he did it as a result of influence exercised by other men, or somehow otherwise as a result of paying too little attention to circumstances, and no offering ought to be made in his behalf. It is incumbent, therefore, upon the Clergyman in any case to investigate the matter accurately, in order to avoid incurring judgment. (St. Nicodemus of the Holy Mountain, in The Rudder 1983 [1957], p. 898)

The central purpose of life is to love and worship God; suicide, including physician-assisted suicide and voluntary euthanasia, involves a purposeful turning away from God. Voluntary euthanasia and physician-assisted suicide sunders one from the Christian recognition of the need for submission to God even unto death.[2]

---

[2]As Engelhardt notes, "There has been no provision to regard such homicide to avoid burdens as other than self-murder" (Engelhardt 2000, p. 330).

Even in secular terms, physician-assisted suicide and voluntary euthanasia raise very serious questions. In the Netherlands, for example, patients 16 years-of-age or older, may provide a written or oral request for medical termination. Physicians are supposed to ensure that the patient's request is well considered and voluntary, and that the patient is enduring unbearable suffering; yet, the meaning of such concerns is deeply ambiguous. Unbearable suffering has included fear of further decline, loss of autonomy and dignity (Regional Euthanasia Review Committees *Euthanasia Code* 2018, revised edition, sections 3.3, 4.2, 4.4). Individuals have requested euthanasia for alcoholism ("My alcoholic brother chose euthanasia" 2016) and mental health problems (Bostaz 2016). If the patient is between the ages of 12 and 16, the Netherlands requires parental or guardian consent for euthanasia. For patients 16 or 17 years of age, the parents or guardians are consulted, but parental consent is not necessary (Regional Euthanasia Review Committees *Euthanasia Code* 2018, revised edition, section 4.2). The willingness to euthanize 16-year-olds on their own authority raises serious questions regarding the ability of adolescents to make well considered or fully voluntary decisions. Here one ought to consider the well documented scientific data demonstrating that the adolescent brain has not yet reached adult development and that adolescent decision-making capacity remains immature, especially in emotionally charged situations. Such physiological and psychological scientific data indicates that it is highly unlikely that adolescents are in a mature enough position to provide adequately informed consent to life ending medical care (see Cherry 2016, pp. 145–167).

Moreover, such laws typically frame the permissibility of requests for medical termination as falling within a limited set of physical or psychological criteria. Yet, in setting out such criteria, laws create a subclass of individuals, i.e., those who fit into the defined categories, whose lives are judged not worth living, degrading the lives of sick, disabled, and dying patients: As David Elliot argues: "to imply that some human lives are less worthwhile due to age or sickness is a ruinous thought, and many think assisted suicide insinuates just this" (Elliot 2018, p. 18). These policies straightforwardly inform specific groups of persons that their lives are the sort of lives that one should reasonably want to end (Elliot 2018). Policy makers, physicians, family members, and others are encouraged to think of such subclasses of persons as individuals who ought to be killed for their own benefit, that these are lives not worth living, and that, perhaps, it would be most appropriate to save such medical resources.

The Canadian Supreme Court, for example, (*Carter v. Canada*, 343–4, para 4) framed its approval of euthanasia and assisted suicide in terms of the patient's own experience of suffering: when "...the person has a grievous and irremediable medical condition (including an illness, disease or disability) that causes enduring suffering that is intolerable to the individual in the circumstances of his or her condition."[3] L. W. Sumner argues broadly in favor of permissible medical termina-

[3] Canada's Supreme Court (Carter v. Canada 2015 SCC 5 [2015] 1 S.C.R. 331) characterized end-of-life decision-making without the possibility of medically assisted death as cruel: "As a result, people who are grievously and irremediably ill cannot seek a physician's assistance in dying and

tion whenever the patient has "…a medical condition (an illness or disability) sufficiently serious to warrant consideration of treatment options that will hasten death" (2017, p. 44). Sumner argues that the condition need not be terminal, but that the patient must experience sufficient suffering that he regards his condition as intolerable, which cannot be otherwise acceptably alleviated (2017, p. 44). Implementing such legislation into medical practice, however, distracts us from recognizing the importance of essential social investment in effective palliative medicine, social and spiritual care for the dying (Eberl 2018).

Autonomy-based arguments that ground justifications for physician-assisted suicide and voluntary euthanasia in what is supposedly individual free choice tend to isolate persons from each other, refusing adequately to appreciate our deep social connectedness, and insisting on individual decisions isolated from the undue influence of others, while also failing to recognize the ways in which patients are socially conditioned to accept such "medical care". It might become tempting to encourage elderly and seriously ill patients towards inexpensive medical termination. After all, "the reasons given to justify assisted suicide do so precisely by characterizing a person's life as irremediably wretched" (Elliot 2018, p. 21). It is worth puzzling about whether the high costs of single payer healthcare in Canada have led to restrictions on palliative, rehabilitative, psychological, and social services, such that medical termination of patients has become a cost saving substitute for quality medical care (Koch 2022). Encouraging patients to die earlier, without the additional costs of extensive treatment for serious disease, saves significant economic resources. Please note, I have no such evidence nor am I making any accusations. However, I challenge readers familiar with euthanasia and assisted suicide practices in Canada, Belgium, Switzerland, the Netherlands, and elsewhere, to take a closer look. Anecdotally, I have been approached by Canadian physicians who report that they are routinely obliged to offer euthanasia as among the options for treatment.

Indeed, Canada's euthanasia statistics are from a certain perspective impressive. Despite having only legalized Medical Assistance in Dying on June 17, 2016, their first annual report documented 13,946 medically assisted deaths from enactment to the end of 2019 (Health Canada 2020, p. 5), by the end of 2020, the total was 21,589 (Health Canada 2021, p. 5). In 2019, approximately 2% of all deaths in Canada were medically assisted; this amounts to 5,631 cases of some form of euthanasia. Health Canada anticipates that these numbers will continue to increase: "The number of cases of MAiD in 2019 represents an increase of 26.1% over 2018 numbers, with all provinces experiencing a steady year over year growth in the number of cases of MAiD since its introduction into law in 2016" (2020, p. 3). In 2020, 2.5% of all deaths in Canada were through some form of MAiD, or 7,595 cases, a 34.2% increase over 2019 (Health Canada 2021, p. 5); 2021 showed an increase to approximately 3.3% of all deaths. As noted, Health Canada reports that "The most commonly cited intolerable physical or psychological suffering reported by patients was

may be condemned to a life of severe and intolerable suffering. A person facing this prospect has two options: she can take her own life prematurely, often by violent or dangerous means, or she can suffer until she dies from natural causes. The choice is cruel" (342–3, para 1).

the loss of ability to engage in meaningful activities (84.9%), followed closely by the loss of ability to perform activities of daily living (81.7%)" (Health Canada 2021, p. 6; see also Health Canada 2022).

A fatal illness, terminal diagnosis, or even significant physical pain, is not always required to be eligible for medical termination. For example, Marc and Eddy Verbessem, 45-year-old deaf, twin brothers, who were slowly going blind were granted euthanasia at Brussels University Hospital. They requested euthanasia on the grounds that they were distraught about going blind and would eventually not be able to see each other (Peace 2013). The Netherlands accepts euthanasia for psychiatric cases, despite the challenges of determining when a psychiatric patient fulfills the requirements of having made a voluntary and well considered decision. In one reported case, a healthy woman was euthanized, whose diagnosis was "prolonged grief disorder", because her husband had died 12 months prior. Among other challenges, the prognosis of psychiatric disorders can be difficult to predict (Kim 2019).

Once one has accepted the centrality of patient autonomy for defining permissible medical choice, it is unclear why the patient must be dying or enduring significant suffering. The patient's subjective experiences, interests, and choices are doing the moral heavy lifting. The value of human life has been reduced to no more than the value that individuals give to their lives. As John Harris concludes, "The point of autonomy, the point of choosing and having the freedom to choose between competing conceptions of how, and indeed why, to live, is simply that it is only thus that our lives become in any real sense our own" (1995, p. 11). Consequently, provided that they are in some sense freely chosen, for secular bioethics, there is little reason to believe that euthanasia and assisted suicide ought only to regard dying patients who are suffering at the end of life. Any competent adult may autonomously choose as part of their medical care to have physicians assist them with their deaths.[4]

Canada's initial law permitting active euthanasia, for example, required only that the individual be at least 18 years-of-age, have the capacity to make health care decisions, give informed consent, and be "Diagnosed with a grievous and irremediable medical condition,' where a person must meet all of the following criteria: serious and incurable illness, disease or disability, advanced state of irreversible decline in capability, intolerable physical or psychological suffering, natural death has become reasonably foreseeable" (Health Canada 2020, 12). In 2020, Bill C-7 removed the requirement for a reasonably foreseeable natural death (Department of Justice 2020). Patients do not need to have a terminal condition. Individuals with a mental illness along with other medical conditions are eligible for active euthanasia.

[4] In Carter v. Canada, the Canadian Supreme Court (369, para 66) specifically cited the importance of patient autonomy as central to their legal reasoning. "An individual's response to a grievous and irremediable medical condition is a matter critical to their dignity and autonomy. The law allows people in this situation to request palliative sedation, refuse artificial nutrition and hydration, or request the removal of life-sustaining medical equipment, but denies them the right to request a physician's assistance in dying. This interferes with their ability to make decisions concerning their bodily integrity and medical care and thus trenches on liberty. And, by leaving people... to endure intolerable suffering, it impinges on their security of the person."

Since the focal point of human life has become autonomous self-determination, secular bioethics, in reflecting such sparse value commitments, rejects traditional religious prohibitions against murder and self-murder, embracing instead personal choices regarding life and death.

Terminology shapes and guides moral discussion. Word choices are not neutral. Language draws out moral understandings and orients users towards ethical or spiritual understandings. These distinctions give expression to and shape the moral life-world. At the same time, shifts in terminology typically signal changes in background moral judgments, axiological understandings, and accepted social practices. For example, referring to euthanasia and physician-assisted suicide as "death with dignity" or "medical assistance in dying", rather than "murder" and "assistance with self-murder," signifies a major shift in taken-for-granted background moral judgments. Describing the practice of deliberately ending the lives of patients as medical treatment or as providing a "dignified" death rhetorically conveys the permissibility of the choice.[5]

For many advocates the preferred language to describe killing patients has shifted away from euthanasia, assisted suicide, or medical termination to the much more conceptually ambiguous euphemism "Medical Assistance in Dying"—a phrase that is strategically designed to assimilate the killing of patients into the everyday practice of medicine. Management of pain and other symptoms, together with medical assistance for nutrition and hydration, are all central elements of typical end-of-life care. Highly vulnerable patients who request or are manipulated into accepting Medical Assistance in Dying may never realize that MAiD encompass active euthanasia. "Medical Assistance in Dying" as a language choice encourages physicians, families, and patients to accept physician-assisted suicide and voluntary active euthanasia as appropriate elements of end-of-life care, alongside of pain management, hospice facilities, and other elements of medical care for the dying.

In short, the significance of taking human life has been deeply blunted, if not fully lost. Death is no longer appreciated as subject to the will of God or as requiring careful spiritual preparation; rather, it is just another event subject to human mastery. Advocates argue that physician-assisted suicide and euthanasia ought to be appreciated as in the patient's best interests because they are thereby able to control the manner and timing of their deaths (Sumner 2017, p. 41). Instead of working to alleviate physical pain or reaching out in love to ameliorate their emotional or psychological anxiety, depression, and despair, family members, friends, and physicians are simply to help patients die. Advocates affirm active medical termination as an appropriate and compassionate solution to avoid what the secular world can only experience as meaningless suffering.

In the end, such practices corrupt physicians and other healthcare workers into killing patients under the guise of doing good. Yet, this is a reality that secular

[5] The ways in which we choose to describe medically assisted death will in turn frame the professional practice of medicine. One might imagine, for example, various professional organizations issuing competing consensus statements in support of quite different medical practices given their competing moral or spiritual commitments.

bioethics cannot comprehend. After all, without God, there is no reason to believe that actively (or passively) killing a person with his consent is morally problematic. The long-term impact on medicine as it continues to accept intentionally ending the lives of patients as part of its professional mores will be significant.

10.4 Some Concluding Thoughts

> For those who renounce Christianity and rebel against it are in their essence of the same image of the same Christ, and such they remain, for until now neither their wisdom nor the ardor of their hearts has been able to create another, higher image of man and his dignity than the image shown of old by Christ. *And, whatever their attempts, the results have been only monstrosities.* (Fyodor Dostoevsky's (1821–1881) reflections in *The Brothers Karamazov*: 1990, 171; emphasis added)

Fyodor Dostoevsky's reflections on the implications of morality without God get right to the heart of the matter once ethics, and thus bioethics, lacks a canonical standard for truth. Violence in the name of equality, social justice, and secular political power has been significant. R.J. Rummel documents some 61,911,000 murdered by the Soviet State; 35,236,000 murdered by the Communist Chinese; 20,946,000 murdered by the Nazis regime; and 2,035,000 murdered by the Khmer Rouge, just to name a few instances (Rummel 1994). The *Victims of Communism Museum* in Washington D.C. documents the killing of more than 100 million people by Communist regimes in their attempts to bring about an equalitarian society and worker's rights (Green 2022). As this chapter argues, the results of rejecting God, of choosing to give meaning and sense to the world ourselves, now includes the dehumanization of entire classes of persons, who are told as a matter of public policy and bioethics that their lives are not worth living. As indicated, such policy works: the number of active medical killings continues to increase. Dostoevsky was correct: the results have very often been monstrous, indeed.

Reference to God, a unique and canonical God's eye perspective, such as Traditional Christianity allows, provides a normative position grounding how all of morality ought to be understood as well as how and why one ought to act on considerations of moral rationality. It provides a perspective necessary to invoke the idea of unconditioned moral truth, for undoing the Gordian knot of moral pluralism. Without God, there is an irresolvable tension between moral rationality and prudential rationality. Why should rational actors follow the requirements of an impartial morality instead of important personal or family-oriented concerns? If moral reality does not come from God, who also acts as the guarantor that the virtuous are rewarded and the vicious punished, it is unclear why one ought to follow an impartial morality once it no longer serves one's own prudential purposes. After all, without God the universe is ultimately meaningless; all acts, even those some deem right, good, virtuous or just, will in the end be lost and forgotten. Traditional Christianity, in contrast, offers the unification of the genesis and justification of morality, while also providing the motivation to be moral.

To return to this volume's framing question: secular reason cannot demonstrate why rationality requires one to be moral or even why it would be prudent to act in accordance with the demands of morality, assuming it could ever definitively justify any specific moral claim. Secular moral epistemological arguments fail (Engelhardt 1996, Chaps. 1–3). If there is no God, then theistic moral arguments are in exactly the same epistemological condition as secular arguments. As a result, insofar as one thinks that there is a necessary connection between the right answer to moral questions and what the law is or ought to be, then without reference to the one true God to establish the correct foundation there is no moral truth upon which properly to establish law, much less any obligation, other than prudential considerations, to follow the legal restrictions imposed. As argued, in such circumstances, at best, only a strictly libertarian state can be justified, in which the actual consent of actual persons is the source of moral authority. However, if the question concerns modern Western democracies, or representative democracies, then the answer turns simply on recognized voting rights. Here, Christians and members of other religions may vote their conscience just like other citizens. Since secular bioethics cannot secure any specific moral content, there is no legitimate reason to limit religious accounts of bioethics from the public forum. Religious citizens may argue for and vote their conscience and worldview just as others do; restricting religious citizens from public debate or democratic voting would constitute abject bigotry.

References

Anscombe, G. E. M. 1958. Modern Moral Philosophy. *Philosophy* 33:1–19.

Bostaz, S. 2016. *Netherlands Sees Sharp Increase in People Choosing Euthanasia Due to "Mental Health Problems"*. Telegraph, May 11. https://www.telegraph.co.uk/news/2016/05/11/netherlands-see-sharp-increase-in-people-choosing-euthanasia-du/.

Carter v. Canada (Attorney General). 2015. *SCC 5 [2015] 1 S.C.R. 331) Online*. Available at: https://scc-csc.lexum.com/scc-csc/scc-csc/en/item/14637/index.do.

Cherry, M.J. 2005/2015. Kidney for Sale by Owner: Human Organs, Transplantation, and the Market. Washington, D.C.: Georgetown University Press.

Cherry, M. J. 2016. *Sex, Family, and the Culture Wars*. New York: Routledge.

Cherry, M. J. 2024. *Bioethics After God: Morality, Culture, and Medicine*. South Bend, IN: University of Notre Dame Press.

Chrysostom, St. J. 1994. Homily I Concerning the Statues to the People of Antioch. In *On the Priesthood, Ascetic Treatises, Select Homilies and Letters, Homilies on the Statues*, edited by P. Schaff, 331–344. Peabody, MA: Hendrickson.

Department of Justice (Canada). 2020. Bill C7: *An Act to Amend the Criminal Code (Medical Assistance in Dying) (justice.gc.ca)*. Available at https://justice.gc.ca/eng/csj-sjc/pl/charter-charte/c7.html.

Dostoevsky, F. 1990. *The Brothers Karamazov*. R. Pevear and L. Volokhonsky (trans.). New York: Farrar, Straus and Giroux.

Downie, J. 2022. From Prohibition to Permission: The Winding Road of Medical Assistance in Dying in Canada. *HEC Forum* 34 (4): 321–354.

Eberl, J. T. 2018. I Am My Brother's Keeper: Communitarian Obligations to the Dying Person. *Christian Bioethics* 24 (1): 38–58.

Elliot, D. 2018. Institutionalizing Inequality: The Physician Criterion of Assisted Suicide. *Christian Bioethics* 24 (1): 17–37.
Engelhardt, H. T., Jr. 1996. *The Foundations of Bioethics*. 2nd ed. New York: Oxford University Press.
Engelhardt, H. T., Jr. 2000. *The Foundations of Christian Bioethics*. Lisse: Swets & Zeitlinger.
Engelhardt, H. T., Jr. 2014. The Orthodox Christian View of Suffering. In *Suffering and Bioethics*, edited by R. Green and N. J. Palpant, 249–261. New York: Oxford University Press.
Frolic, A., and P. Miller. 2022. Implementation of Medical Assistance in Dying as Organizational Ethics Challenge: A Method of Engagement for Building Trust, Keeping Peace and Transforming Practice. *HEC Forum* 34 (4): 371–390.
Frolic, A., and A. Oliphant. 2022. Introducing Medical Assistance in Dying in Canada: Lessons on Pragmatic Ethics and the Implementation of a Morally Contested Practice. *HEC Forum* 34 (4): 307–319.
Goodall, David. 2018. *David Goodall: Scientist, 104, Ends his Life in Switzerland*. BBC News May 10. https://www.bbc.com/news/world-europe-44069885.
Green, D. 2022. *Victims of Communism Museum Review: A Red Reminder*. Wall Street Journal, July 06.
Habermas, J. 2008. *Between Naturalism and Religion*. Ciaran Cronin (trans.). Cambridge: Polity Press.
Harris, J. 1995. Euthanasia and the Value of Life. In *Euthanasia Examined: Ethical, Clinical and Legal Perspectives*, edited by J. Keown, 6–22. Cambridge: Cambridge University Press.
Health Canada. 2020. *First Annual Report on Medical Assistance in Dying in Canada 2019*. https://www.canada.ca/en/health-canada/services/publications/health-system-services/annual-report-medical-assistance-dying-2019.html.
Health Canada. 2021. *Second Annual Report on Medical Assistance in Dying in Canada 2020*. https://www.canada.ca/en/health-canada/services/publications/health-system-services/annual-report-medical-assistance-dying-2020.html.
Health Canada. 2022. *Third Annual Report on Medical Assistance in Dying in Canada 2021*. https://www.canada.ca/en/health-canada/services/publications/health-system-services/annual-report-medical-assistance-dying-2021.html.
Health Canada. 2023. *Fourth Annual Report on Medical Assistance in Dying in Canada 2022*. Online: https://www.canada.ca/en/health-canada/services/publications/health-system-services/annual-report-medical-assistance-dying-2022.html.
Kelly, G. 1957. *Medico-Moral Problems*. Dublin: Clonmore & Reynolds, Ltd.
Kim, S. 2019. *How Dutch Law Got a Little Too Comfortable with Euthanasia*. The Atlantic, June 8. Available at https://www.theatlantic.com/ideas/archive/2019/06/noa-pothoven-and-dutch-euthanasia-system/591262/.
Koch, T. 2022. A Sceptic's Report: Canada's Five Years Experiment with Medical Termination (MAiD). *HEC Forum* 35 (4): 357–369.
Mark the Ascetic. 1979. On Those Who Think That They Are Made Righteous by Works. In *The Philokalia: The Complete Text*, edited by Sts. Nikodimos and Makarios and Trans. G. E. H. Palmer, P. Sherrard, and K. Ware, 125–46. Boston: Faber and Faber.
Milton, J. 1997 [1667]. *Paradise Lost*. London. Judy Boss (transcriber.) Renascence Editions. https://scholarsbank.uoregon.edu/xmlui/bitstream/handle/1794/767/lost.pdf?sequence=1.
My alcoholic brother chose euthanasia. 2016. BBC News December 01. https://www.bbc.com/news/av/world-europe-38166185.
Peace, W. J. 2013. *Euthanasia in Belgium: The Untold Story*. The Hastings Center, Bioethics Forum Essay. https://www.thehastingscenter.org/euthanasia-in-belgium-the-untold-story/.
Pellegrino, E. D. 2001. Physician-Assisted Suicide and Euthanasia: Rebuttals of Rebuttals—The Moral Prohibition Remains. *Journal of Medicine and Philosophy* 26 (1): 93–100.
Pennock, L. 2022. *Exclusive: 'I Don't Want to Die, But I Don't Want to be Homeless': Canadian Man, 65, has a Doctor's Approval for Euthanasia Despite Admitting Becoming POOR is a Main Reason He's Applying to Die. Daily Mail 9 December 2022*. https://www.dailymail.

co.uk/news/article-11516989/Canadian-man-doctors-approval-euthanasia-despite-admitting-POVERTY-main-factor.html.

Regional Euthanasia Review Committees (The Netherlands). 2018. *Euthanasia Code*, rev. ed. http://euthanasiecommissie.nl and https://english.euthanasiecommissie.nl.

Rummel, R. J. 1994. *Death by Government*. New Brunswick, NJ: Transaction.

St. Nicodemus of the Holy Mountain. 1983 [1957]. *The Rudder*. Chicago: The Orthodox Christian Educational Society.

Sumner, L. W. 2017. *Physician-Assisted Death: What Everyone Needs to Know*. New York: Oxford University Press.

Wheithman, P. J. 1999. Of Assisted Suicide and "The Philosopher's Brief". *Ethics* 109 (3): 548–578.

Chapter 11
Circumcision as a Critical Test Case for the Value of Religious Bioethics: A Close Examination of the Writings of Rabbi Elliot N. Dorff

Max Buckler and Brian D. Earp

Abstract This chapter examines the role of religion in bioethics through the lens of male circumcision, focusing on the work of Rabbi Elliot N. Dorff, a prominent religious bioethicist. The chapter analyses Dorff's distinctively religious approach to bioethics, grounded in the Jewish principle that "God owns the body" (GOTB). While this precept leads Dorff to generally oppose non-voluntary, non-therapeutic body modifications, it concomitantly suggests an exception for penile circumcision, justified on religious grounds. The presence of this exception creates tensions with Dorff's broader ethical framework, particularly regarding bodily integrity and autonomy. The chapter also explores the implications of Dorff's stance for other forms of child genital cutting, such as those practiced by some Islamic communities (e.g., "ritual pricking" of the vulva of young girls). The authors contend that Dorff faces a dilemma: opposing such practices would conflict with his commitment to religious pluralism, while supporting them would seem to contradict his own religious precepts. How a distinctively religious approach to bioethics can best handle such a dilemma remains an open question. Ultimately, the authors call for a more robust engagement from religious bioethicists like Dorff on contentious issues such as child genital cutting. They argue that the topic serves as a crucial test case for religious ethics in bioethics, challenging thinkers to apply their principles consistently across different practices and belief systems.

M. Buckler (✉)
Independent Scholar, Hong Kong, China

B. D. Earp
Centre for Biomedical Ethics, Yong Loo Lin School of Medicine, National University of Singapore, Singapore, Singapore

K. R. Smith (ed.), *Bioethics and Religion*, The International Library of Bioethics 116, https://doi.org/10.1007/978-3-032-33538-8_11

11.1 Introduction

Should religion be permitted a seat at the table in bioethics? For this chapter, we have been asked to consider how the topic of circumcision might inform a response.

The simple answer is: Yes, of course, religion (and especially religious individuals) should have a seat at the table of bioethics, where bioethics is understood to be a deliberative, theoretical and practical inquiry into the normative dimensions of biomedicine, biotechnology, and related areas. Since religious people, practices, and beliefs are often implicated in these normative dimensions (the case of circumcision, being sometimes a religious and sometimes a medical procedure, is a good example of this; see Reis 2021), it is both necessary and desirable for bioethics and bioethicists to engage directly with the topic of religion, and with religious people.

Of course, there are different views about how best to "do" bioethics, or what rules should govern bioethical debate and deliberation, whether about religious topics or seemingly more secular ones. But whatever these "rules of engagement" are, or should be, we would simply maintain that they must apply equally to all participants. So, while religion (and especially religious individuals) should certainly have a seat at the table of bioethics, the seat should be of the same general size and make as any other. In short, there should be no *special* seats for religious bioethicists, or for non-religious ones, or anyone else.

In all cases, debaters should be clear about their premises and show how their conclusions follow. And if they aim to be persuasive, it will usually be prudent to pick premises that are widely shared among the interlocutors. Alternatively, it may be useful to show how *different* premises can—when combined with certain plausible auxiliary assumptions—get you to the same destination. Anyone can make a good or a bad argument. So, for example, secular thinkers sometimes propose weak or objectionable arguments about matters of religion or bioethics, and religious thinkers sometimes propose strong and compelling ones. Other times, it's the other way around. And sometimes, as we'll see, religious and non-religious philosophers reach similar conclusions about what should be done, or what some policy should be, albeit via different pathways.

As one of us has argued, "we all have certain meta-ethical commitments (whether implicit or explicit, religious or otherwise) and we all have to try to convince those we disagree with that our meta-ethical commitments make more sense than theirs do, or do a better job of explaining a shared moral intuition, or whatever. That's just 'doing philosophy'" (Earp 2015, p. 865; responding to Biggar 2015). In other words, we *all* should strive to come to the table with clear argumentation, consistent positions, and a readiness to receive and respond to critique.

In this chapter, we will examine the issue of nontherapeutic childhood circumcision (medically unnecessary removal of part or all of the penile foreskin of a child who is too young to consent) from a philosophical and bioethical perspective. Predominately, we will be concerned with the notion of a child's right to bodily integrity (RBI) and how this might best be understood.

When writing in our own voice, we will approach the topic from a secular bioethical perspective: that is, we will not appeal to any particular (i.e., particular to some community rather than widely shared) religious or metaphysical viewpoint to justify claims or arguments. In other words, we will do our best to proceed on the basis of public reason (Rawls 1997; Van Howe 2013a). However, it will also be important to know how a religious bioethicist might think about the notion of a child's RBI in relation to circumcision. For it would shed light on a deeper question, namely: In the absence of medical necessity, when, if ever, are the religious beliefs or motivations of a child's parents or guardians sufficient to justify *any kind* of surgical intervention into a child's sexual anatomy (Van Howe 2013b; Sarajlic 2014; Möller 2017; Chambers 2018; for an alternative perspective, see Jacobs and Arora 2015)?

As a test case for this line of inquiry, our focus will be on the writings and speeches of noted bioethicist Rabbi Elliot N. Dorff, to see how his distinctively *religious* approach to bioethics might contribute to longstanding debates about child genital cutting. We have chosen to focus on Rabbi Dorff because, as we are tasked with exploring the relevance of religion to bioethics, few better exemplify the intersection.

As a rabbi and Jewish educator, Dorff has been hailed as the foremost bioethicist of the Jewish Conservative movement, and a guiding figure of Jewish bioethics writ large. Yet, Dorff has equally been sought by important secular ethics groups and initiatives (indeed, up to the national level) to provide an authoritative religious perspective on a breadth of moral issues. Of course, his positions, like the positions of any rabbi or educator, cannot be read as *the* definitive characterization of Jewish ethics (especially as regards the particular character of the Jewish religion, which is an interpretive tradition). However, he has often been viewed by secular bodies interested in incorporating religious perspectives into their work as a leading authority on Jewish religious law. According to Louis E. Newman, founding president of The Society of Jewish Ethics:

> Elliot's reputation extends far beyond the Jewish community. His standing as one of the country's leading Jewish ethicists has earned him invitations to address several national commissions and bioethics task forces, including the National Bioethics Advisory Commission, where he has testified on human cloning and stem cell research; the Surgeon General's Commission on Responsible Sexual Behavior; and the National Human Resources Protections Advisory Commission. He also served, in 1993, on the Ethics Committee of Hillary Rodham Clinton's Health Care Task Force. In addition, Elliot has participated in several other collaborative projects through the American Association for the Advancement of Science and the Carnegie Foundation. In all these contexts, Elliot offers a distinctively Jewish voice, yet one capable of speaking to a diverse audience. And in each of these settings, he has earned a reputation for the depth of his learning, the respectful way he listens to others, and the incisive way he frames the key issues. It is no wonder that whenever people are looking for a Jewish authority on major moral issues, Elliot's name rises to the top of the list. (Newman and Dorff 2024, p. xiii)

Indeed, speaking generally, we, too, regard Dorff as a worthy illustration of how a religious ethicist can fruitfully participate in, and contribute to, bioethics as a discipline. He comes to the "table" with thoroughly researched and well-argued positions, and a wealth of knowledge about the strengths and weaknesses of competing moral systems as they relate to contemporary issues. Importantly, Dorff is also not shy about critiquing his own religious tradition; addressing moral pitfalls of various religious practices; reassessing positions based on new information; and identifying areas for—and pathways toward—moral progress within his religious community. The potential for such progress, as we will show below, may be attributed in large part to Dorff's high valuation of vigorous debate.

So, what is his position on circumcision? Seemingly, this would be a central issue on which Dorff, given his influence, approach, and specialized training, would be especially well-suited to contribute. After all, circumcision of the penis—a matter of particular importance in traditional Jewish practice while also being customary in Islam and incidentally practiced within some Christian communities—falls directly at the intersection of religion, biomedicine, health, and sexuality.

Curiously however, we are left to speculate on Dorff's precise conception of the ethics of circumcision and of the child's right to bodily integrity (RBI), as he has said surprisingly little about either. For example, in Dorff's chapter on "Beginning of Life" issues in *The Oxford Handbook of Jewish Ethics and Morality* (Dorff and Crane 2016), although he takes positions on multiple relevant issues including abortion and reproductive rights, genetic testing, sexual health, and more, there is not a single mention of circumcision. The subject is likewise mostly absent (mentioned only in passing, and not engaged with as a bioethical concern) from Dorff's seminal three-volume series, *Matters of Life and Death: a Jewish approach to modern medical ethics* (1998), *To Do the Right and the Good: a Jewish approach to modern social ethics* (2002), and, *Love Your Neighbor and Yourself: a Jewish approach to modern personal ethics* (2003), which are considered by many as canonical works of contemporary Jewish ethics.

This is, in our view, a striking omission, not only due to the significance of circumcision in traditional Judaism, but also due to its being a bioethically contested US American cultural practice (Bedzow 2020; Munzer 2018). Without knowing Dorff's position on this critical issue, or how it relates to his position on other, seemingly analogous practices, such as "ritual pricking" of the vulva for religious reasons (discussed below), we may be missing a crucial insight into the role of religion in bioethical debate. In what follows, then, we take a deeper look at the nature and strength of Rabbi Dorff's distinctively religious approach to bioethics, hoping that this may serve as a window into the value of religious approaches to bioethics more generally.

11.2 Framing our Discussion: The Right to Bodily Integrity

Opponents of non-therapeutic, non-voluntary circumcision of the penis (NNC, we'll say) typically argue that the practice conflicts with the child's right to bodily integrity (RBI) (e.g., Van Howe et al. 1999). How exactly this right should be formulated is a matter of ongoing debate (Mazor 2013; Viens 2017; Townsend 2023;

Earp 2022a; Pugh 2023; Mazor 2024), but arguments in this vein typically highlight the following features of NNC which it shares with other genital cutting practices widely viewed as problematic:

(a) It targets, and concentrates surgical risk on, the sexual anatomy of a non-consenting person;
(b) It removes healthy, sensitive tissue from—and thus permanently alters—a person's genitals (i.e., their so-called "private" parts) without their permission;
(c) It does these things without medical necessity

On some views, then, *whatever* the child's RBI consists of, it is the sort of thing that is violated—or at least, infringed[1]—by any practice that has these features (BCBI 2019, 2024; see also Higashi 2023). In addition to so-called "normalization" surgeries on the genitals of infants born with intersex traits, which are becoming increasingly controversial (Carpenter 2018; Ziemińska 2020; Muschialli et al. 2024), the most obvious set of practices that share these features with NNC are the ones that fall under heading of 'female genital mutilation' or 'FGM' as defined by the World Health Organization (WHO): for example, what is called FGM Type 1a (cutting or removal of tissue from the clitoral hood, leaving the labia and clitoral glans/body intact); FGM Type 2a (cutting or removal of tissue from the inner labia, without cutting or modifying other parts of the vulva); or FGM Type 4, which "includes all other harmful procedures to the female genitalia for non-medical purposes, e.g., pricking, piercing, incising, scraping and cauterizing the genital area" (WHO 2023).

Importantly, the WHO definition and typology for 'FGM' does not apply solely to highly invasive forms of ritualistic cutting or modification of the vulva, such as infibulation, but also to forms that are, on any view, much *less* invasive than NNC of the penis, and which are, moreover, often done in the same families for similar reasons (O'Neill et al. 2020; Shweder 2022; Buckler 2024; Earp et al. 2024). Thus, it includes a practice known in English as 'ritual pricking' or 'nicking' (*khatna* in Arabic; FGM Type 1a or 4 on the WHO typology), in which a ceremonial drop of blood is drawn from a young girl's vulva, as we explore later in the chapter. According to the WHO, *all* such female genital practices, being both medically unnecessary and non-consensual, violate the child's RBI—that is, *irrespective* of the anticipated harm-level, *irrespective* of parental motivation, *irrespective* of whether the practice has been medicalized, and *irrespective* of any claimed benefits, whether physical or psychosocial (Van Eekert et al. 2024). If that is correct, it might seem to follow that NNC must likewise violate a child's RBI, *whatever* that right consists of, because it appears to meet the same basic criteria (Davis 2003; Van Howe 2011).

Of course, not everyone agrees with the WHO's position on 'ritual pricking'—much less the proposed extension of this position to NNC. As proponents of (or defenders of toleration for) ritual pricking argue, the child's RBI is not, in fact,

[1] Philosopher Joseph Mazor, for example, suggests that the child's RBI may permissibly be infringed upon (and therefore not actually violated) in cases when a competing interest or other right may be argued to take priority (see Mazor 2019).

violated by the practice, because it is insufficiently *harmful* (Arora and Jacobs 2016; Duivenbode and Padela 2019; Porat 2021; Shweder 2022). Setting aside how 'harm' is to be understood in such cases (Van Howe 2011; Earp and Darby 2017), such an analysis implies that a child's RBI is only violated by practices that, perhaps among other things, meet some minimal threshold of harmfulness. More generally, then, it is ostensibly an *empirical* matter (once we agree what harm is, how to measure it, and what the actual threshold should be) whether any given practice is in fact harmful enough to violate a child's RBI.

On the other hand, the RBI might be an *autonomy*-based right, or at least a right whose 'job' is to protect (future) autonomous decision-making (for discussions, see Earp 2019; Bagattini 2015; Munzer 2018; Pugh 2023). Thus, it might entail that certain types of decisions—for example, decisions about whether non-therapeutic surgical risk should be concentrated on one's sexual anatomy—should be left to the affected individual to make, if and when they are able to make it. According to this interpretation, if a person is *capable* of consenting to circumcision or ritual pricking, but fails to do so, the right is violated by the practice, *irrespective* of any anticipated empirical consequences. And similarly, this view suggests, if a person is *not (yet) capable* of consenting to a medically unnecessary genital cutting procedure, to impose the practice anyway would be to violate their *future* right to decide for themselves (i.e., it would violate the right in advance) (Munzer 2018; BCBI 2024; see also Darby 2013; and more generally, Feinberg 2015).

We will not aim to settle such matters in this chapter (though we have argued elsewhere for something like the latter position: see Buckler 2023; see also Earp and Steinfeld 2018; Earp 2019). Our point is only that the bioethicist whose work we are discussing, Rabbi Elliot N. Dorff, has not clarified his position either way.

Suppose he *does* think NNC (and perhaps also ritual pricking) at least infringes the child's RBI, but that the state should permit such infringement if it is done for religious reasons. If so, there are various ways to make such an argument (e.g., Jacobs and Arora 2017), and it would be useful to see them made. On the other hand, suppose he does *not* think NNC infringes, much less violates, the child's RBI. Well, then, it might be that he accepts a broadly harm-based formulation of the right (along with the necessary conceptual and empirical premises). Or perhaps he thinks that *ritual pricking* violates the right, but NNC does not—a possibility we explore below. Either way, it would be illuminating to know: What is Rabbi Dorff's position?

The answer to this question matters. And it is a key example of why we need religious bioethicists "at the table" to think our way through it. Suppose we are responsible for setting up a task force to determine policy recommendations related to the child's RBI. Our job might be to determine under what, if any, conditions, a clinician should be permitted to cut a small child's genitalia for non-therapeutic reasons. Now, who might we select for such a group? Would we benefit from Rabbi Dorff's perspective?

11.3 Dorff's Jewish Ethics

Although Dorff does not directly analyze NNC or discuss ritual pricking in his works on ethics, we do have some clues about his general approach to conceptualizing bodily integrity. Dorff commonly describes the core foundation of Jewish bioethics as tied to the theological conception that "God owns the body" (hereafter, "GOTB"). Here is how Dorff explains this precept in his chapter on "Beginning of Life" issues in *The Oxford Handbook*:

> A number of Judaism's fundamental convictions affect Jewish approaches to the moral issues at the beginning of life.
>
> 1. The body belongs to God. Unlike American secular ethics, in which each person's body belongs to him or herself, Jewish classical texts assert that God, as Creator of the universe, owns everything in it, including our bodies. Therefore God can and does make certain demands of us as to how we use our bodies, demands articulated in Jewish law. It is as if we were renting an apartment: we have fair use of the apartment during our lease—and, in the biological analogue, during our lease on life—but we do not have the right to destroy the apartment (commit suicide) or harm it unnecessarily, because it is not ours. (Dorff and Crane 2016, p. 313)

Now, secular liberal ethicists may balk at this suggestion, as it appears on its face to be—far from a common starting point for discussion—a rather dogmatic illiberal creed. For one, it implies that it is morally appropriate for humans to be treated as property (albeit belonging to a transcendent owner), and thus, on Dorff's theological explanation, not entitled to independent, individual rights over their own bodies.[2] The specifics of God's "demands" on us may also be worrying, insofar as some may be at odds with otherwise widely accepted contemporary ethical norms. But perhaps even more concerning, to the secularist, is the implication that this principle applies not only as a guide for how individuals should treat their own bodies, but also in terms of how empowered agents (for example parents and medical professionals) ought to treat vulnerable individuals at the beginning of life (such as infants and small children) who may or may not grow up to share the adults' overarching metaphysic (Earp 2022b).

However, it is crucial to understand that Dorff is not a religious fundamentalist or legal positivist. He does not argue that the biblical commandments—no matter how explicitly stated in the text—are beyond moral evolution through human interpretation and critique. Indeed, as Dorff's co-editor of the *Oxford Handbook*, and fellow scholar of Jewish ethics, Jonathan K. Crane, notes:

> One could take Dorff to task, [by] saying that he (sometimes) holds a wrong faith...Dorff evades this critique by turning not to claims of belief or theology but to hermeneutics. Hence he argues that norms bequeathed to the living must be continuously and communally re-evaluated in light of contemporary circumstances and sensibilities and thus avoid

[2] Dorff suggests that Jewish ethics operates through enforcement of *duties* (or commandments) rather than *rights*. He claims that this is a valuable mode through which to pursue ethics as it adds an active rather than passive impetus to behave morally (Newman and Dorff 2024, see section beginning on page 31).

> becoming "wrong" in the strong sense. This process therefore juxtaposes moral sensibilities and the Judaic textual tradition because, according to Dorff, they are neither coextensive nor coterminous. (Crane 2009, p. 151)

Accordingly, Dorff acknowledges, "with the possible exception of some right-wing Orthodox groups, all modern Jews see the world through Enlightenment glasses, where the individual is the fundamental reality" (Dorff 2007, p. 33). Dorff also strongly defends the separation of church and state. For example, he has said that "Jewish law, [however] interpreted, should not be the basis of American law and neither should the stance of any other religion" (Congregation Brith Shalom 2022). And so, Dorff cedes priority to civil law to govern over all religions (more on this below).

One of the most helpful explanations Dorff offers for engagement with his work is a reference he makes to Thomas Jefferson's conception of "the public square," in which Dorff claims, "the model is that you come together to determine public policy bringing with you your convictions, whatever they are. And the presumption is that through debate, very vigorous debate, that you would ultimately come to a much wiser public policy than if people came to that discussion pretending that they didn't have [their personal] convictions" (Glazer Institute 2013). Therefore, upon our assessment, Dorff uses the principle of GOTB, not as a dogma, but more so as a particularistic "framework" or perhaps a "state of mind" which he brings to ethics discourse. He uses this framework to argue for a religiously grounded general ethic but without an invincible conviction that any specific religious moral or traditional practice will remain accepted merely on the merit that his religion commands it.

This explains why Dorff rarely if ever uses the theological notion of GOTB to defend controversial scriptural edicts in the modern context. Much to the contrary, Dorff's positions have a marked tendency to align with the liberal mainstream, leading Dorff to take opposing conclusions to many of his contemporaries in Orthodox branches of Judaism. For example, Dorff interprets the Jewish religious tradition as functionally compatible with reproductive freedom even though many Orthodox authorities disagree (Dorff and Crane 2016; Eisenberg 2004). Dorff supports gender equality in civil and religious matters despite thousands of years of codified religious patriarchy in Judaism (Kimmel 2001). He supports sexual equality/freedom, and opposes domestic violence (Dorff 1995), notwithstanding certain passages in the bible that some could argue fail appropriately to condemn instances of domestic violence (as such behavior would be interpreted today) (Karman 2022).

More recently, Dorff has reassessed one of his previously most conservative positions, on so-called 'medical aid in dying' (MAID); while formerly categorically opposed, he has softened over the last few decades, in line with advancements in civil policy and pressure from religious peers (Dorff 2020). There is perhaps no better example of Dorff's openness to liberalism than his co-authoring of an influential 2006 *responsum* (rabbinic ruling) which permits the ordination of openly gay clergy (Dorff 2006).

Thus, Dorff's hermeneutical approach makes space for a seemingly modern liberal perspective. On the one hand, this reflects a commendable pragmatism on

Dorff's part, as he pursues a theologically satisfying mode through which to balance competing commitments between parochial and universalist values. Yet from a secular viewpoint, it is unclear how to apply Dorff's central meta-ethic that "God owns the body" to practical issues in bioethics, since (at least as presented by Dorff) we can never be certain whether God's instructions (and by extension, contemporary religious practices) are righteous exactly as written (or as currently practiced), or outdated and deserving of reinterpretation, including through appropriate internal and external criticism. And nowhere is this more challenging than when it comes to the issue of the liberal values of bodily integrity and autonomy, and their interaction with the seemingly illiberal practice of NNC (Shweder 2009).

As previously noted, Dorff has not, to our knowledge, committed to any substantive treatment of this intersection, especially not in his writings on bioethics (we discuss a partial exception shortly). He has, however, made some brief comments about religious circumcision in other texts, and we will use those to analyze his bioethics positions involving other issues that clearly bear on bodily integrity and autonomy. As we will show, this analysis reveals some puzzles and contradictions.

11.4 Dorff on Religious Circumcision

While the topic of circumcision is largely absent from Dorff's works on Jewish bioethics, he has touched on the subject in other books[3] focused on religious practice and social action. A brief but interesting commentary on the topic can be found in his book *The Way into Tikkun Olam (repairing the world)* (2007), where he describes circumcision as one of several Jewish "parental duties." As a part of the segment, Dorff touches on contemporary internal Jewish criticism of circumcision:

> In our own day, some Jews have questioned the wisdom of circumcising boys. The relevant medical associations have wavered on its medical advisability, recommending it during the last several decades of the twentieth century and now saying that its medical risks and benefits are roughly equivalent, making circumcision medically optional.
>
> The Jewish tradition, however, never advocated circumcision for medical reasons. Its purpose was instead religious. (Dorff 2007, p. 210)

To begin with, it is interesting that Dorff's immediate response to internal Jewish critique of circumcision is to reference the wavering pronouncements of non-religious medical associations (though he does not cite any in particular, we assume he means US groups; see Earp and Shaw 2017 for context). At first glance, this might make it seem that Dorff thinks the permissibility of religious circumcision

[3] For example, see the section on circumcision in *Jewish Choices, Jewish Voices: Body* (2008), edited by Dorff. While he does not take a position on the controversies himself, he includes opposing perspectives from two other Jewish authors. Feminist Harry Brod expresses regret over circumcising his son, while urologist and *mohel* Samuel Kunin defends newborn circumcision in his essay, "Circumcision: The Medical Imperative Accompanying the Religious Imperative" (Dorff and Newman 2008).

might depend in some way on medical pronouncements. It would be illuminating for Dorff to expand on this question. Although he clarifies that medical reasons were not traditionally grounds for positively "advocating" circumcision within Judaism (for example, due to claims about "health benefits"), it is less clear what he would say if relevant medical associations advised *against* NNC, as some European associations now do (see Svoboda and Van Howe 2013; for a recent comparison of European and British policy, see Lempert et al. 2023). Suppose a majority of such associations one day claimed to have established that the risks or harms of NNC definitively outweigh any health-related benefits. Would *this* then be relevant to the ethical status of a practice whose purpose is understood to be religious? There might be an important asymmetry here. Although Dorff seems to be saying that circumcision's being "recommended" on health grounds would not necessarily be a pertinent reason to advocate circumcision as a matter of Jewish tradition, he does elsewhere describe Jewish principles, including GOTB, as disallowing the imposition of bodily *harms* in certain cases (as we discuss in more detail below). This might seem to suggest that a finding of harm by relevant medical associations would indeed be relevant to the intra-religious debate. On the other hand, if God "owns" the body, it might instead be thought that God could make an exception to God's own general rule against harming the body, namely by allowing, or even requiring, certain harmful bodily practices due to their special religious significance. Either way, it would be helpful to know more about Dorff's overall position on the relevance, or possible lack of relevance, of secular medical findings about the "benefits versus risks" of NNC, *given* his view that its purpose is essentially religious, rather than something health-related or medical.

In the meantime, however, we will interrogate his claim that NNC can be seen as "medically optional."

This portrayal of NNC as "medically optional" anticipates the posture of the American Academy of Pediatrics in its 2012 policy on NNC (AAP 2012). The policy, which was allowed to expire in 2017, concluded that "the preventative health benefits [of NNC] outweigh the risks of the procedure" (AAP 2012, 585), but that the benefits were not strong enough to recommend circumcision as a routine procedure for all boys. However, the latter qualification was often dropped in media coverage of the policy, such that the headline claim of benefits outweighing risks seemed to shape much of the public's understanding. Many therefore initially interpreted the 2012 policy as a positive *endorsement* of the practice on medical grounds. Upon this reading of the term, "medically optional" could be taken as a "soft" medical recommendation without a mandate.

However, subsequent comments from one of the main contributors to the AAP policy lend themselves to a different interpretation. These comments suggest that the policy was not solely designed to dispassionately settle an evidence-based medical question. Instead, it was intended to support, and make scientifically defensible, a contested assumption of parental authority: namely, the authority to choose circumcision for *non-therapeutic reasons*, against a tide of increasing rights-based criticisms of the practice when applied to non-consenting children (Freedman 2016). Thus, although it was couched in the seemingly neutral language of medical benefits and risks, the policy's main conclusion rested in part on non-scientific values and motivations. As the lead urologist for the policy task force, Dr. Andrew Freedman, explained in a 2016 clarification piece, these values included "religion, culture, aesthetic preference, familial identity, and personal experience." For the task force members, "protecting this option [to circumcise newborns for such non-medical reasons] was not an idle concern at a time when there are serious efforts in both the United States and Europe to ban the procedure outright" (Freedman 2016, p. 2). On this interpretation, the term "medically optional" might be better understood as an allowance or justification, but not a positive medical recommendation. Consistent with this interpretation, in a recent interview with one of the current authors, Freedman stated: "the best analogy is that the AAP guidelines are a 'permission slip' for those who want to circumcise their children so that society cannot say they are bad parents or outlaw the practice," while fellow task force member Dr Douglas Diekema, himself a prominent bioethicist, stated: "It was really a legal question for me. My feeling was that there was not sufficient data to suggest that this is a procedure that should be outlawed, particularly given that there were multiple religious communities for whom this was an important practice. But I also didn't think paediatricians should be recommending it" (Buckler, 2025, p. 2).

For full disclosure, one of us has previously critiqued the 2012 AAP policy on both scientific and ethical grounds, arguing that the central claim about benefits outweighing risks was not logically or medically justified (Earp and Shaw 2017). Other critics reached similar conclusions (Frisch et al. 2013; Svoboda and Van Howe 2013; see also Van Howe 2018). We won't rehash those criticisms here. Rather, we want to highlight a more general conceptual issue. When religious and purported medical justifications for NNC are presented together, it can become unclear which is doing the normative work. If the practice is said to confer health benefits, one need never ask whether the religious rationale would be regarded as sufficient in the absence of such benefits. The medical rationale can always be invoked, even if it is not the fundamental justification. Dorff's position is valuable in this respect because he largely avoids this ambiguity. From his quoted remarks, he clearly understands Jewish religious circumcision as being performed for non-therapeutic reasons. Given this, and his apparent acceptance of the practice, what follows? Seemingly, that Dorff believes that under certain conditions, permanent surgical body modification can be justified without medical necessity, without needing even to invoke a medical reason, and without the consent of the person affected. This is the standard against which we will compare Dorff's other positions on body modification below.

As to the non-medical rationales that Dorff himself gives for circumcision (a subject which is an open matter of debate amongst other Jewish religious authorities), he writes:

> First, by literally inscribing a sign of God's covenant with Israel on a man's [sic] flesh, circumcision symbolizes that this covenant between God and the People Israel is immutable. Compare this to sealing a document in contrast to signing it: the seal changes the very surface of the paper, making it virtually impossible to change the document.
>
> Second, because circumcision changes the surface of a male's generative organ, it symbolizes that this covenant is to last from generation to generation...

> Third, circumcision has implications for *tikkun olam* as well. In graphically tying Jews to the covenant, circumcision symbolizes Jews' commitment to Judaism's moral duties at least as much as to its rites. These include, of course, the duties [regarding] speech, the poor, captives, the sick, and one's parents and children. (Dorff 2007, p. 210–211)

In summary, Dorff produces the following religious rationales for circumcision: (1) a symbolic rationale (which includes irreversibility as an important feature, thus requiring a permanent change in morphology rather than mere surface-level marks on the skin), (2) a familial and communal commitment and relation, and (3) an assignment of particularistic *morality,* for which Dorff suggests that altering a child's "generative organ" will play a role. (We say "child's" here rather than "male's"—or as Dorff says inexplicably in the first line, "man's"—because the symbol in question is not traditionally inscribed into the flesh of a "man" or a generic "male," but rather, into the flesh of a newborn child; moreover, this fact is absolutely central to current debates about the ethical permissibility of NNC, so should not be rhetorically occluded or glossed over).

While we are skeptical that any such reasons are sufficient to morally justify the performance of a non-therapeutic surgery on a young child's genitals (or for that matter on the genitals of any person who is not capable of consenting), our skepticism is not the focus of this chapter. Rather, what we are examining at present is whether Dorff can defend these arguments as permitting a *type* of practice he would endorse as a matter of principle, or if he instead views them as uniquely appropriate to newborn penile circumcision (perhaps especially as it is performed within traditional Judaism). We suspect the latter is more likely, as when we examine Dorff's other positions on non-therapeutic body modifications, it seems there are some values in conflict. These conflicts, we believe, stem from the fact that non-voluntary body modification isn't just an unusual or questionable practice by post-Enlightenment standards, but also by the standards of traditional Judaism (as understood and expounded upon by Dorff) as well.

11.5 An Internal Religious Exception as Much as an External One?

Although circumcision is viewed as foundational within most contemporary sects of Judaism, it is also exceptional in that the practice is not emblematic of (indeed, it is contrary to) Judaism's *general* religious ethic on body modification. In line with the biblical commandment in Leviticus 19:28, the Torah forbids ritual cuts to the body (e.g., for mourning the dead) or inscribing permanent marks, and Jewish scholars throughout history have consistently interpreted this as prohibiting practices like tattooing and other aesthetic body modification. Some even apply this precept as requiring the body to remain intact after death (Dorff 1998, 238) (which is why some Orthodox authorities prohibit autopsies and cadaver organ donation; see Rosner 1971). Therefore, the value of leaving the body unmodified is not only familiar to Jewish religious authorities and traditional communities, it is an accepted value within those communities and is typically given significant normative weight by the associated authorities.

Thus (in keeping with his religious commitments), in most cases Rabbi Dorff presents the general Jewish ethic as *negatively* predisposed to non-therapeutic body modification, such as cosmetic surgery, tattoos, and piercings, even for consenting adults. For example, Dorff writes of tattoos that the rabbis of the Talmud believed "that putting permanent marks on one's skin violated the divine image in which we were all created" (Dorff 1998, p. 268).

However, in keeping with his apparent liberal leanings, we find that in almost all cases Dorff ultimately allows for (consenting, adult) individuals to choose for themselves. Despite his evident disapproval of such practices and endorsement of the religious prohibition as the ideal, he usually reasons that when a person modifies *themselves*, it can be tolerated religiously if their individual subjective values can be perceived as healthy in some meaningful way.

Consider the following two passages, in which Dorff assesses the ethics of cosmetic surgery and genital piercings:

> [Dorff on cosmetic surgery]
>
> Whereas therapeutic surgery seeks to reestablish what existed before and was damaged, cosmetic surgery done for aesthetic reasons seeks instead to improve the body a person inherited…
>
> …The proper response to such [aesthetic] pressure, however, is not to undergo the risks of cosmetic surgery to resemble others but rather to take nonsurgical steps to gain a better estimation of oneself…
>
> …In line with this view, one might think that cosmetic surgery for aesthetic reasons alone should be prohibited on theological, if not legal, grounds, for it is effectively saying that what God gave us is not good enough. As such, it is an act of ingratitude toward God, a lack of appropriate appreciation of the body God gave us. (Dorff 1998, p. 271)
>
> [Dorff on genital piercing]
>
> In general, Jewish law prohibits injuring ourselves or others. Body piercing escapes that prohibition because it is done in the name of adorning, rather than injuring, oneself. If the point of piercing the body is to add to one's beauty, though, one cannot but wonder how many people besides the person so adorned are supposed to see the pierced genital areas and thus appreciate them as a mark of beauty. If it is only the person whose body is pierced and that person's spouse, well and good. The Jewish value of modesty, though, demands that others should not be shown those areas of the body or the piercing or rings attached to them. (Dorff 1998, 270)

The first thing to note is that, as mentioned, Dorff generally frowns upon non-therapeutic body modification. And yet, this conflicts with NNC in a straightforward way: NNC is itself a non-therapeutic body modification, but is religiously mandated while other analogous body modifications are religiously forbidden. But in a deeper sense, this also speaks to a conflicting *application of values*. Recall the positive religious rationales Dorff applied to NNC: an irreversible aesthetic transformation of the body, the binding of communal affiliation, and the motivation (indeed, commandment) to act in a certain moral manner. The high moral value Dorff places on NNC stands in contrast to the amorality or immorality he assigns to other body modifications. Circumcision is celebrated precisely for permanently altering the natural—or as some might prefer, "God-given"—body while other practices with similar physical outcomes demonstrate "a lack of appreciation of the body God gave us" (Dorff 1998, 271).

This suggests that the positive values applied to NNC are so exceptional that they are not even fit to serve as an *exemplary model* for other body modifications. Dorff derides movies and television for unduly influencing people to modify themselves. But if influencing (let alone "commanding") surgical body modification is a bad thing for a cultural movement to do, then what does that mean when religions (which are, after all, *also* cultural movements; Brusa and Barilan 2009) do the same? And how can Dorff's disvalue of surgical modification for the sake of "resembling others" be reconciled with NNC: a practice which is intended to mark male membership in the group collective, such that all are expected—indeed, most are forced—to "resemble" each other in a very specific way?

Dorff's comments on genital piercing are especially curious in this regard, as they seem to demonstrate consternation towards others who value or perform alterations of the *genitals* in particular. This is surprising since it could reasonably be assumed that if one regards a non-therapeutic genital modification as foundational to one's own belief system, one would be at least sympathetic toward others who also practice genital modifications, even if their preferred modification or specific motives may be different than one's own. However, in the passage above, Dorff appears perplexed by the idea that anyone would want to interfere with their intact or "God-given" genitalia.

More broadly, Dorff does not appear to entertain the notion that one who pierces their own genitals (let alone someone else's) might *also* be invested in some significant spiritual or communal objectives, such as marking their "generative organ" for the purposes of intergenerational continuity, or, perhaps, out of fealty to someone or something the individual believes owns their body—remarks that plausibly apply, not only to Jewish penile circumcision, but also to the Islamic practice of "ritual pricking" as we will soon examine. Indeed, Dorff even states that religious virtue "demands" that such behavior be kept as private as possible, clearly interpreting genitals as an intimate domain (a traditional as well as a contemporary concern; see Bettcher 2023; Buckler et al. 2023), rather than as a part of the body to be marked or displayed in the context of public (e.g., community-building) behavior.

Now, it may be argued that since Dorff ultimately condones or tolerates most common body modification practices (i.e., he doesn't advocate their prohibition), there is no real conflict with NNC.[4] But it must be understood that the basis for Dorff's begrudging attitude of allowance toward these practices is his respect for *individual subjectivity and autonomy*. As he writes: "in the end such procedures are

[4] Other religious authorities have noted that even Orthodox Jewish law can make space for certain body modifications and piercings (such as ear piercing and cosmetic procedures such as rhinoplasty) if they have become commonplace in society. For a discussion, see filmmaker Eliyahu Ungar-Sargon in conversation with Rabbi Asher Lopatin (Ungar-Sargon 2018, 19:10). But there is no indication that these may be performed non-consensually, and abstention from body modification remains the normative value, while modifications require special justification.

permitted in the name of augmenting a person's own sense of self-worth and, hence, mental health" (Dorff 1998, 272). Personal choice is what ultimately overrides the religious prohibition for Dorff. In other words, when it really matters to his position in the public square, Dorff is—effectively, if not very enthusiastically—a supporter of bodily autonomy.

Yet in all such cases in his works on medical ethics of which we are aware, Dorff approaches the ethics of aesthetic body modification solely from the angle of permissibility for a competent decision maker to alter their *own* body. Nowhere does Dorff apply this reasoning to the notion that an adult may consider using their subjective aesthetics or conceptions of mental health—much less, their religious beliefs—to justify tattooing or surgically altering *another* person's body for non-therapeutic reasons without their consent. In fact, Dorff specifically observes that the importance of ritual circumcision within Judaism *overrides* a general Jewish value of personal privacy. He writes:

> There are ... several ways in which Jewish practices ... promote intrusion. For example ... friends and family are expected to celebrate the *brit milah* (ritual circumcision) of a newborn boy with his parents, even though that is surgery on his most private parts. Here social needs to express appropriate emotions and to articulate traditional Jewish understandings of a life-cycle event ... outweigh the needs for privacy. These are exceptions, though, to the general rules [dictating norms of privacy]. (Dorf 2003, 43)

Thus, as stated earlier, circumcision is an exceptional practice, even *within* Dorff's own religious framework.

Now, this is a simple enough inconsistency to reconcile internally, where the primary impetus to follow God's commandments may be taken to justify exceptional practices, if God (who works in mysterious ways) endorses them; but it is a more difficult matter to explain when it comes to the public square. It is hard for one to defend a practice to outsiders when in most other similar cases they would strongly disapprove. Dorff may point to Genesis 17, Leviticus 12, and other sources to explain Jewish circumcision. However, if we asked Dorff for the Judaism-informed bioethical position on the permissibility of a practice involving cutting or excision of *other* healthy genital tissues, or the amputation of (part of) some *other* healthy bodily organ, we expect that his answer would follow much as it does in the *Oxford Handbook*: that because neither our bodies—nor our children's bodies—are our own, we should not have the "right" to alter or harm them.

Now, consider what this would mean for our hypothetical task force on the child's RBI. It seems to us that because of circumcision's exceptionality in Judaism, the Jewish principle of GOTB as presented by Dorff would not be helpful to our analysis, as it is used to explain both the exception *and* the rule.

Yet this leaves us with a major remaining question with practical consequences for the debate: How does Dorff approach *other* religious genital cutting practices, where the "social needs" of the community could likewise be invoked as "outweighing" norms around privacy or bodily integrity? If Dorff may claim that it is permissible to cut a child's body based on the belief that God has commanded it, or favors it, would he be willing to accept this very same argument when presented by a follower of a different religion? We explore this question in the following section.

11.6 Does Dorff's Position Obligate Him to Oppose Genital Cutting or Alteration if Pursued by Other Religions?

So, what about the followers of other religions? Should they also be permitted to cut their children's genitals? This question is not merely theoretical, as there are several other types of religiously motivated child genital cutting practices globally. However, many Western states have enacted legislation prohibiting other forms of medically unnecessary child genital cutting, including all types affecting children with female-typical genitalia, regardless of the religious beliefs or intent of the parents/community, and no matter how minor or "symbolic" the given procedure may be (see, e.g., the US federal STOP FGM Act of 2020, passed into law as Public Law No: 116-309 in 2021; for context, see Earp 2020).

Indeed, a recent string of high-profile legal cases in the USA and Australia have highlighted this issue, with controversy centering on the Dawoodi Bohra community (a Shi'a Muslim sect with roots in South Asia), whose female genital cutting practice is allegedly constrained to a mitigated "symbolic" ritual nick, or prick (Rogers 2019; Bootwala 2019a, 2019b, 2019c). Members of the Bohra community who support the practice have argued that there is a double standard at play, since penile circumcision—which they also practice—is currently permitted in Western countries with or without a religious rationale, and even when clearly both medically unnecessary and non-consensual, yet is substantially more physically invasive than their female variant (Zakir 2016; DBWRF 2018).

While Dorff has not commented on these controversies,[5] our reading of his positions leaves us unsure what he would recommend, since we could equally imagine him reaching one of two diametrically opposed positions:

1. Dorff could *condemn* other types of child genital cutting, including by supporting protections for children, since his own religious code ordinarily compels him to do so.
2. He could *condone* other forms of child genital cutting, since he usually endorses religious freedom as a civic right, and equal treatment of different religions, even if his own internal religious code would prohibit other types of religious practice.

While it is difficult to speculate which of these positions Dorff would adopt, we believe that either one risks undermining the coherence of his general philosophy. No matter which course of argument he prefers, he will have to betray a major value. Allowing other types of child genital cutting rituals would be inconsistent with the norms around bodily privacy he usually argues for; yet prohibiting them

[5] Dorff provides the following explanation as to why Judaism itself does not include a female genital cutting variant to accompany ritual circumcision of the penis: "The same symbolism was not needed in removing a girl's clitoris, presumably because from ancient times to the modern era the man was the head of the house, and therefore making men part of the covenant throughout all generations immediately made all the female members of their families part of the covenant as well" (Dorff 2007, 211).

would seem to offend the value of religious pluralism and equal respect for different religions.

Consider position #1. As established earlier, Dorff claims that a fundamental conviction in Judaism is that the body belongs to God (GOTB) and we are therefore forbidden from making permanent, non-therapeutic changes to the body or otherwise causing the body "unnecessary" harm. This precept extends to a seemingly generalized moral opposition to aesthetic body modification as described above. As noted, Dorff allows for certain exceptions to this prohibition in the case of consenting adults choosing for themselves, but in the case of *non-consensual* genital cutting it cannot be attached to uniquely personal reasons. It would therefore seem, at least upon a first look, that a position *opposing* child genital cutting (CGC) would be the most natural outgrowth of Dorff's usual arguments.

In fact, there is another major reason why Dorff could be expected to oppose CGC, which we have not yet touched upon directly, which is the potential of such practices to be experienced as acts of sexual trespass or violence. Many self-described victims of religiously motivated genital cutting compare the experience to sexual assault—or other violation of sexual boundaries—including in situations where the cutting itself is supposedly 'minor' (at least relatively so), as with ritual pricking (Kumar 2021). Consider the experience of Saleha Paatwala, a member of the Dawoodi Bohra community, who was taken by her grandmother, when she was 7 years old, to undergo the prescribed ritual for girls in her community. Referring to the unfamiliar woman who carried out the procedure at her grandmother's behest, she states:

> She asked me to lie down and, uh, this very thought gave me goosebumps all over my body. [This] woman started pulling down my underwear. And that whole idea...it was really scary. She took it out and now she spread my legs, grabbed the blade, and cut something between my legs. It was definitely painful. [But] it was more embarrassing because a lady whom I did not know saw my [private] area at that point of time, and she did not see but she also cut... (Paatwala 2018).

When considered in this light, the stakes of ethical debates around CGC move beyond an assessment of the practices considered as body *modifications* and expand to include an assessment of them as *actions* taken by adults toward children's genitalia: actions whose symbolic significance—as interpreted by the child, whether in the moment, or upon later reflection—cannot be fixed by the adults in advance (Buckler et al. 2023). In other words, as one of us has argued elsewhere, debates about the bodily *state* produced by a non-consensual genital procedure (e.g., whether this state reflects, or amounts to, a net benefit or harm) miss the more basic consideration that such a procedure involves, necessarily and in the first instance, intrusion into the child's sexual embodiment: that is, a part of the body commonly considered to be especially intimate or private, such that a person can be violated by "mere" non-consensual touching of, or even looking at, their genitalia (see Box 11.1), at least outside of certain rare medical emergencies (Buckler 2023).

Considered in the abstract, this is a point with which, we expect, Rabbi Dorff would be inclined to strongly agree. After all, he, too, argues that the genitals should be seen as a distinctively private and protected site, including from a Jewish religious perspective. See, in particular, his fascinating *responsum* on "family violence"

(Dorff 1995). On the impermissibility of interfering with another person's genitals (even for an otherwise righteous, explicitly non-sexual cause), Dorff writes,

> The Torah makes this exceedingly clear: 'If two men get into a fight with each other, and the wife of one comes up to save her husband from his antagonist and puts out her hand and seizes him by his genitals, you shall cut off her hand; show no pity [Deut. 25:11-12]:' Despite the special justification the woman had for shaming her husband's assailant, the Torah demands drastic steps in retribution for the degradation she caused (Dorff 1995, 790).

Thus, although we earlier learned of Dorff's position that ritual circumcision promotes "intrusion" through friends and family participating in "surgery on [a boy's] most private parts" (Dorff 2003, 43), Dorff otherwise concludes that, "We are humiliated when we are sexually abused—even just touched in our private parts against our will—for we feel that our sense of self has been invaded, that our honor has been compromised in the most fundamental way possible" (Dorff 1995, 790). And Dorff is especially protective of children. He writes: "Indeed, in light of the extensive damage it causes to the future ability of the child to cope with life, without too much exaggeration I would say that, in the case of children, sexual abuse is akin to murder" (Dorff 1995, 791).

What does this imply for Dorff's likely position on ritual pricking in the Bohra community? On the principles considered so far, strong opposition would seem the most natural conclusion. Dorff's account of God's ownership of the body establishes a general presumption against bodily injury, while the exceptions he allows for non-therapeutic body modification ordinarily depend on the autonomous choice of the person whose body is altered. Moreover, as we have just seen, he treats even the non-consensual touching (much less cutting) of another person's genitals as a profound invasion of personal boundaries, even if there is no sexual intent. It would therefore be challenging for him to defend an accommodation of the practice, whether by means of a legal excuse or other type of exception (see Merkel and Putzke 2013), without explaining why these considerations should yield in this particular case.

Box 11.1 Can medically unnecessary genital procedures, even if performed with good intentions, be experienced as a violation of sexual boundaries? The question of timing

Whether a medically unnecessary genital procedure is experienced or interpreted by the individual as a type of trespass on their sexual boundaries could depend, some might argue, on the timing. For example, one could argue that in the case of ritual pricking, insofar as it is carried out while the child is fully conscious and aware of her genitalia as private anatomy, there is a heightened risk that the act will be experienced or interpreted as a sexual violation, as compared to penile circumcision as performed in infancy, that is, before the child is likely forming any consciously-retrievable memories, which is when it is traditionally practiced within Judaism.

However, first, it must be noted that boys in many communities, including some Muslim communities, are also often circumcised under coercive conditions while conscious and sexually self-aware: that is, as older children, rather

(continued)

Box 11.1 (continued)

than in infancy (Lee 2006; WHO 2008; Anwer et al. 2017; Rizalar et al. 2017; Yavuz and Kobya Bulut 2024). Moreover, there is evidence—from a range of cultural settings—that an unknown proportion of men (and some transgender women) who underwent NNC in fancy or childhood do feel a sense of sexual violation or other significant distress in relation to the procedure, sometimes upon later reflection (Bossio and Pukall 2018; Remennick 2022; Hammond et al. 2023; Uberoi et al. 2023; Tao 2024). With respect to Jewish practice, although the newborn period is when the vast majority of circumcisions are performed, circumcision of older boys does sometimes occur, for example, in the case of adolescent males who convert to Judaism (see also Remennick 2022 for a different example). There is also the Jewish ritual practice of *hatafat dam brit*—which, similar to the Bohra practice for girls—involves pricking or piercing the sexual organ to draw a drop of blood. *Hatafat dam brit* is more commonly performed on older children (typically, already-circumcised males whose circumcisions were considered ritually invalid); and it is also experienced as a sexual boundary violation by some who undergo it. As one man recounted (quoted in Buckler 2024; see that paper for further examples and discussion),

> I was...old enough to be worried when Dr. Greenbaum said, "Joshua, please lower your pants." I began bawling as I inched my tightie-whities down my thighs, revealing my teenage [penis]. The doctor grasped my offending member [and] pricked my bunched skin and squeezed until a blood dollop oozed from the minute hole...On the rabbi's command, I stripped and stepped into the mikveh [ritual bath]...Instead of pride, I felt shame: my rabbi had seen me naked before my eighth-grade girlfriend.

This is not to say that ritual genital procedures performed on infants or newborns have no implications for sexual boundaries. For example, suppose the Bohra community—like various other communities that traditionally practice ritual female, as well as male, genital cutting—decided to change the timing of the practice to infancy (in some communities, this is a response to the practice becoming criminalized; see Earp et al. 2024), perhaps based on the reasoning that "she won't remember it." We think it is unlikely this would mollify those who regard the practice as a serious trespass. This is because, first, in the case of unambiguous violations (e.g., sexual molestation), it is typically considered morally *worse*, not better, if the victim is unconscious or a very small child, irrespective of whether the child later learns about what happened.

And second, as we and others have suggested elsewhere, while it is true that "very young children, including infants, do not (yet) [have] a sense of their genitalia as 'intimate' anatomy...most will come to acquire such [a] sense. If they learn, therefore, that prior to their ability to exercise [an] essential boundary-setting right, their 'intimate' anatomy was *already* cut or altered for reasons other than medical necessity, they may reasonably come to conclude that (what should have been) an exceptionally personal choice about their sexual embodiment has been usurped" (BCBI 2024, p. 8).

On the other hand, consider the potential for taking position #2, and the challenges this might raise for Dorff's approach to bioethics: namely, arguing for the *permissibility* of other types of CGC. Dorff ostensibly supports total equality in civil law, and claims there is a Jewish religious basis for this position, as he writes, "[In] thirty-six passages, by the rabbis' count, the Bible boldly proclaims equality in law between Jew and alien" (Dorff 1998, 272). Thus, in most cases we would expect Dorff to support religious freedoms equally and without discrimination within a pluralistic society, including in relation to practices which are prohibited by the particularistic requirements of Judaism. And, as we've seen, Dorff is well-equipped to navigate around many biblical prohibitions where they conflict with contemporary values. Thus, it is not inconceivable that Dorff could come up with a flexible approach (as he does with personal cosmetic surgery) to other religious practices out of respect for value pluralism.[6]

Setting aside the aforementioned potential for these practices to be experienced as violations of sexual boundaries (Box 11.1), the relatively minimal invasiveness of certain types of female genital cutting (FGC) countervails oft-raised concerns over detrimental physical changes, which is a common knee-jerk objection to such cutting (see generally, Catania et al. 2007). Some of these practices are indeed so "minor" that they can hardly be said to modify the body at all, and it is therefore argued by some of the groups who practice such rites that they are not harmful to the body or its natural function (see above). In fact, such a position was once supported by the AAP, as well, when it attempted a policy aimed at permitting ritual pricking or a similar procedure in order to offer a "medically optional" practice (or rather, "a medicalized option") to relevant religious communities (AAP 2010; for discussion, see Van Howe 2011).[7]

Yet, regardless of the potential for these practices to be mitigated in their physical impact, we also point out that Dorff has placed religious value in *significant* morphological change in the body for penile circumcision, a procedure that removes, in the form that currently predominates within Jewish practice, roughly one-third to one-half of the motile skin system of the penis (Taylor et al. 1996; see also Fahmy 2020; Cepeda-Emiliani et al. 2023). (Recall how Dorff explained that "sealing a document" would be greater than just "signing it.") So, it seems reasonable that Dorff could also potentially support some acceptable *range* of genital modifications (ranging from the most superficial to perhaps some types of relatively minor modifications, for example, of the inner labia), if for no other purpose than to stay consistent in his reasoning.

However, the taking of either of these positions—condemn or condone—would have challenging implications for Dorff's central ethic. Per Dorff's commitment to equality in civic law, it could be viewed as hypocritical for him to oppose FGC in its most minor forms, as whatever protections can be offered to girls should apply equally to boys (and perhaps also to children with intersex traits, whether assigned

[6] Indeed, such a position has been forcefully defended by the anthropologist Richard Shweder (2009, 2022).

[7] Paradoxically, this position struggles to gain traction because of the very fact that pricking/nick is *so* minor it cannot be attached to possible health benefits *à la* circumcision. See Buckler (2024).

female or male; see BCBI 2024; DeLaet et al. 2024). Yet taking the second position—*condoning other forms of CGC*—might be perceived as equally problematic for Dorff, particularly as it relates to the question of this volume: namely, Should religion play a role in bioethics?

This is because Dorff cites God's ownership of the body as a fundamental religious conviction (GOTB), guiding both his personal ethic and his distinctively religious approach to bioethics. When, therefore, is it relevant to bioethical debate? As we have seen, it appears that, in practice, Dorff's religious conviction most often yields to contemporary secular arguments based on liberal values, autonomy-promoting stances on the body, and the requirements of public reason. It may be wondered, then, what the principle of GOTB really adds to bioethical analysis? After all, insofar as the principle seems to conflict with secular liberal conclusions, God, for the most part, does not seem to get His way.[8]

Although there are no easy answers here, we believe that such questions must be raised. With Dorff, we agree that even the most contentious issues in bioethics require vigorous debate—including both religious and secular perspectives—in the public square.

11.7 Conclusion: An Invitation to Meet in the Public Square

As stated at the beginning of this chapter, we welcome good-faith religious engagement in the field of bioethics as much as any other good-faith engagement, so long as the same argumentative rules apply across the board. Accordingly, we respect Rabbi Dorff's immense efforts in the field of ethics, and we have focused on his work here precisely because it can be so fruitfully put into conversation with that of others. As Dorff writes, "Although God may be omniscient, no human being is, so the best way for us to attain truth, goodness, and wisdom is to pool our intellectual and moral resources in serious conversation" (Newman and Dorff 2024, 100).

We would therefore invite Rabbi Dorff to weigh in on the topic of religious NNC of the penis, and its relation to the child's RBI, particularly when considering other forms of child genital cutting and modification done for religious reasons, such as ritual 'pricking' of the vulva as performed in some Muslim communities. Given the significance of NNC to traditional Judaism, and the increasing degree of controversy surrounding the practice, the time for public discussion including both secular and religious viewpoints on the topic has surely come. A scholar of Dorff's pedigree should be expected to contribute to any such discussion.

When we consider arguments from any philosopher—religious or not—we expect their positions to apply consistently, and we expect them to engage forthrightly in difficult challenges to their convictions, especially when they apply to contested practices. This is not to say exceptions to general principles can never be

[8] We are using "his" self-consciously here; it is how God is typically gendered in Jewish (and Christian, etc.) scripture.

made when weighing competing values, but when a philosopher proposes an exception (and hopes to convince us of its validity), they should be able to explain what precise conditions must be met for the exception to apply. Consequently, wherever else those conditions are met, the exception should also be available to others equally (Shweder 2017; Ahmadu 2017; Ahmadu and Kamau 2022).

For these reasons, we believe the issue of child genital cutting is a crucial test case for Dorff, and for religious bioethics more generally. We invite him to share his views on this matter.

References

AAP. 2012. Circumcision Policy Statement. *Pediatrics* 130 (3): 585–586.

Ahmadu, F. S. 2017. Equality, Not Special Protection: Multiculturalism, Feminism, and Female Circumcision in Western Liberal Democracies. In *Universalism Without Uniformity: Explorations in Mind and Culture*, edited by J. L. Cassaniti and U. Menon, 214–236. Chicago: University of Chicago Press.

Ahmadu, F. S., and T. Kamau. 2022. Dr Tatu Kamau vs The Attorney General and Others: Problems and Prospects in Kenya's 2021 High Court Ruling to Uphold the Prohibition of Female Genital Mutilation Act 2011 - A Reply to 'The prosecution of Dawoodi Bohra Women' by Richard Shweder. *Global Discourse* 12 (1): 29–46.

American Academy of Pediatrics [AAP] Committee on Bioethics. 2010. Ritual Genital Cutting of Female Minors. *Pediatrics* 125 (5): 1088–1093.

Anwer, A. W., L. Samad, S. Iftikhar, and N. Baig-Ansari. 2017. Reported Male Circumcision Practices in a Muslim-Majority Setting. *BioMed Research International* 2017 (1): 4957348.

Arora, K. S., and A. J. Jacobs. 2016. Female Genital Alteration: A Compromise Solution. *Journal of Medical Ethics* 42 (3): 148–154.

Bagattini, A. 2015. Male Circumcision and Children's Interests. In *The Nature of Children's Well-Being: Theory and Practice*, edited by A. Bagattini and C. Macleod, 113–128. Dordrecht: Springer.

Bedzow, I. 2020. Male Circumcision As An Example of Bioethics as (Immaterial) Social Ethics. *Ethics, Medicine and Public Health* 14:100519.

Bettcher, T. M. 2023. Phenomenology, Agency, and Rape. *Feminist Philosophy Quarterly* 9 (2): 1–6.

Biggar, N. 2015. Why Religion Deserves a Place in Secular Medicine. *Journal of Medical Ethics* 41 (3): 229–233.

Bootwala, Y. 2019a. A Review of Female Genital Cutting (FGC) in the Dawoodi Bohra Community: Part 1—FGC Terminology, Western Genital Cutting Practices, Southeast Asian Type Ia and Type IV FGC Practices. *Current Sexual Health Reports* 11:212–219.

Bootwala, Y. 2019b. A Review of Female Genital Cutting (FGC) in the Dawoodi Bohra Community: Part 2—Bohra Culture, FGC Practices in Dawoodi Bohras, and Pertinent Legal Cases. *Current Sexual Health Reports* 11:220–227.

Bootwala, Y. 2019c. A Review of Female Genital Cutting in the Dawoodi Bohra Community: Part 3—The Historical, Anthropological and Religious Underpinnings of FGC in the Dawoodi Bohras. *Current Sexual Health Reports* 11:228–235.

Bossio, J. A., and C. F. Pukall. 2018. Attitude Toward One's Circumcision Status is More Important Than Actual Circumcision Status for Men's Body Image and Sexual Functioning. *Archives of Sexual Behavior* 47:771–781.

Brusa, M., and Y. M. Barilan. 2009. Cultural Circumcision in EU Public Hospitals: An Ethical Discussion. *Bioethics* 23 (8): 470–482.

Brussels Collaboration on Bodily Integrity (BCBI). 2019. Medically Unnecessary Genital Cutting and the Rights of the Child: Moving Toward Consensus. *The American Journal of Bioethics* 19 (10): 17–28.

Brussels Collaboration on Bodily Integrity (BCBI). 2024. Genital Modifications in Prepubescent Minors: When May Clinicians Ethically Proceed? *The American Journal of Bioethics*, online ahead of print at https://www.tandfonline.com/doi/abs/10.1080/15265161.2024.2353823.

Buckler, M. 2023. The Ethics of Child Genital Cutting. When Does a Violation Occur? Comments on 'Defending an Inclusive Right to Genital and Bodily Integrity for Children' by Dr. Kate Goldie Townsend. *International Journal of Impotence Research* 35 (1): 31–34. https://doi.org/10.1038/s41443-022-00611-2.

Buckler, M. 2024. The Smallest Cut: The Ethics and (Surprising) Implications of Hatafat Dam Brit for the Ongoing Genital Cutting Debate. *Kennedy Institute of Ethics Journal*. Pre-print: https://www.researchgate.net/profile/Max-Buckler/publication/382430537.

Buckler, M. 2025. As controversies mount, circumcision policies need a rethink. Journal of Medical Ethics, online first. https://doi.org/10.1136/jme-2025-111357

Buckler, M., L. Bruce, and B. D. Earp. 2023. From Intimate Exams to Ritual Nicking: Interpreting Nonconsensual Medicalized Genital Procedures as Sexual Boundary Violations. *Current Sexual Health Reports* 15 (4): 291–300.

Carpenter, M. 2018. Intersex Variations, Human Rights, and the International Classification of Diseases. *Health and Human Rights* 20 (2): 205–214.

Catania, L., O. Abdulcadir, V. Puppo, J. B. Verde, J. Abdulcadir, and D. Abdulcadir. 2007. Pleasure and Orgasm in Women with Female Genital Mutilation/Cutting (FGM/C). *The Journal of Sexual Medicine* 4 (6): 1666–1678.

Cepeda-Emiliani, A., M. Gándara-Cortés, M. Otero-Alén, H. García, J. Suárez-Quintanilla, T. García-Caballero, et al. 2023. Immunohistological Study of the Density and Distribution of Human Penile Neural Tissue: Gradient Hypothesis. *International Journal of Impotence Research* 35 (3): 286–305.

Chambers, C. 2018. Reasonable Disagreement and the Neutralist Dilemma: Abortion and Circumcision in Matthew Kramer's Liberalism with Excellence. *The American Journal of Jurisprudence* 63 (1): 9–32.

Congregation Brith Shalom, dir. 2022. *Rabbi Elliot Dorff A Jewish Perspective on Abortion Rights*. https://www.youtube.com/watch?v=hS7nIpOwUSQ.

Crane, J. K. 2009. *Rhetoric of Modern Jewish Ethics*. Available online: https://tspace.library.utoronto.ca/bitstream/1807/17748/1/Crane_Jonathan_K_200906_PhD_thesis.pdf.pdf.

Darby, R. J. 2013. The Child's Right to an Open Future: Is the Principle Applicable to Non-Therapeutic Circumcision? *Journal of Medical Ethics* 39 (7): 463–468.

Davis, D. S. 2003. Cultural Bias in Responses to Male and Female Genital Surgeries. *The American Journal of Bioethics* 3 (2): W5.

DBWRF. 2018. *Official Statement About the Implement Application Filed by DBWRF*. Dadoodi Bohra Women's Association for Religious Freedom. https://dbwrf.org/blog_details/15.

DeLaet, D. L., B. D. Earp, and E. Mills. 2024. Which Children Have Rights? The Child's Right to Bodily Integrity and Protection Gaps for Children With Intersex Traits Under International and National Laws. *Amicus Curiae* 5 (3): 448–473.

Dorff, E. N. 1995. *Family Violence*. Rabbinical Assembly. https://www.rabbinicalassembly.org/sites/default/files/assets/public/halakhah/teshuvot/19912000/dorff_violence.pdf.

Dorff, E. N. 1998. *Matters of Life and Death*. Jewish Publication Society.

Dorff, E. N. 2002. *To Do the Right and the Good: A Jewish Approach to Modern Social Ethics*. Jewish Publication Society.

Dorff, E. N. 2003. *Love Your Neighbor and Yourself: A Jewish Approach to Modern Personal Ethics*. Jewish Publication Society.

Dorff, E. N. 2006. *Homosexuality, Human Dignity, and Halakhah: A Combined Responsum for the Committee on Jewish Law and Standards*. Rabbinical Assembly. https://www.rabbinicalassembly.org/sites/default/files/public/halakhah/teshuvot/20052010/dorff_nevins_reisner_dignity.pdf.

Dorff, E. N. 2007. *The Way Into Tikkun Olam (Repairing the World)*. Jewish Lights Publishing.
Dorff, E. N. 2020. *Assisted Suicide/Aid in Dying Reconsidered*. Rabbinical Assembly. https://www.rabbinicalassembly.org/sites/default/files/2021-04/Assisted%20Suicide%20Revisited%20final.pdf.
Dorff, E. N., and J. K. Crane. 2016. *The Oxford Handbook of Jewish Ethics and Morality*. Oxford University Press.
Dorff, E. N., and L. E. Newman. 2008. *Jewish Choices, Jewish Voices: Body*. Jewish Publication Society.
Duivenbode, R., and A. I. Padela. 2019. Female Genital Cutting (FGC) and the Cultural Boundaries of Medical Practice. *The American Journal of Bioethics* 19 (3): 3–6.
Earp, B. D. 2015. Does Religion Deserve a Place in Secular Medicine? *Journal of Medical Ethics* 41 (11): 865–866. https://doi.org/10.1136/medethics-2015-102917.
Earp, B. D. 2019. The Child's Right to Bodily Integrity. In *Ethics and the Contemporary World*, edited by D. Edmonds, 217–235. Routledge.
Earp, B. D. 2020. Why was the US Ban on 'Female Genital Mutilation' Ruled Unconstitutional, and What Does This Have to do With Male Circumcision? *Ethics, Medicine and Public Health* 15 (1): 100533.
Earp, B. D. 2022a. The Ethics of Circumcision. In *The Rowman & Littlefield Handbook of Bioethics*, edited by E. Di Nucci, J.-Y. Lee, and I. Wagner, 296–317. Lanham, Maryland: Rowman & Littlefield.
Earp, B. D. 2022b. Against Legalising Female 'Circumcision' of Minors: A Reply to 'The Prosecution of Dawoodi Bohra Women' by Richard Shweder. *Global Discourse* 12 (1): 47–76.
Earp, B. D., and R. Darby. 2017. Circumcision, Sexual Experience, and Harm. *University of Pennsylvania Journal of International Law* 37 (2): 1–56.
Earp, B. D., and D. M. Shaw. 2017. Cultural Bias in American Medicine: The Case of Infant Male Circumcision. *Journal of Pediatric Ethics* 1 (1): 8–26.
Earp, B. D., and R. Steinfeld. 2018. Genital Autonomy and Sexual Well-being. *Current Sexual Health Reports* 10:7–17. https://doi.org/10.1007/s11930-018-0141-x.
Earp, B. D., J. Abdulcadir, and A. Shahvisi. 2024. What Counts as Mutilation—And Who Should Decide? Disrupting Dominant Discourses on Genital Cutting and Modification. *Culture, Health and Sexuality*, online ahead of print at https://www.tandfonline.com/doi/full/10.1080/13691058.2024.2388438.
Eisenberg, Daniel. 2004. *Abortion in Jewish Law*. Aish.Com (blog). May 30, 2004. https://aish.com/abortion-in-jewish-law/.
Fahmy, M. A. B. 2020. *Normal and Abnormal Prepuce*. Dordrecht: Springer.
Feinberg, J. 2015. The child's right to an open future. In *Justice, politics, and the family, 145–158. Routledge.*
Freedman, A. L. 2016. The Circumcision Debate: Beyond Benefits and Risks. *Pediatrics* 137 (5): e20160594. https://doi.org/10.1542/peds.2016-0594.
Frisch, M., Y. Aigrain, V. Barauskas, R. Bjarnason, S. A. Boddy, P. Czauderna, et al. 2013. Cultural Bias in the AAP's 2012 Technical Report and Policy Statement on Male Circumcision. *Pediatrics* 131 (4): 796–800.
Glazer Institute, dir. 2013. *Rabbi Elliot Dorff "Religion in the Public Square: Jewish Civic Engagement"*. https://www.youtube.com/watch?v=jNhXHnzS_Ns.
Hammond, T., L. M. Sardi, W. A. Jellison, R. McAllister, B. Snyder, and M. A. Fahmy. 2023. Foreskin Restorers: Insights Into Motivations, Successes, Challenges, and Experiences with Medical and Mental Health Professionals–An Abridged Summary of Key Findings. *International Journal of Impotence Research* 35 (3): 309–322.
Higashi, Y. 2023. Autonomy, bodily Integrity and Male Genital Cutting. In *Female Genital Mutilation/Cutting: Global Zero Tolerance Policy and Diverse Responses from African and Asian Local Communities*, 159–177. Singapore: Springer Nature Singapore.
Jacobs, A. J., and K. S. Arora. 2015. Ritual Male Infant Circumcision and Human Rights. *The American Journal of Bioethics* 15 (2): 30–39.

Jacobs, A. J., and K. S. Arora. 2017. Punishment of Minor Female Genital Ritual Procedures: Is the Perfect the Enemy of the Good? *Developing World Bioethics* 17 (2): 134–140.
Karman, Y. 2022. A tale of two Tamars: Domestic violence in the Hebrew bible. *Verbum et Ecclesia* 43 (2442): 1–5.
Kimmel, M. S. 2001. Circumcision: The Kindest Un-Cut: Feminism, Judaism, and My Son's Foreskin. *Tikkun* 16 (3): 43–48.
Kumar, R. 2021. *In India's Bohra Community, a Battle Over Genital Mutilation*. Undark. March 10. https://undark.org/2021/03/10/india-battle-over-female-genital-mutilation/.
Lee, R. B. 2006. Filipino Experience of Ritual Male Circumcision: Knowledge and Insights for Anti-Circumcision Advocacy. *Culture, Health and Sexuality* 8 (3): 225–234.
Lempert, A., J. Chegwidden, R. Steinfeld, and B. D. Earp. 2023. Non-Therapeutic Penile Circumcision of Minors: Current Controversies in UK Law and Medical Ethics. *Clinical Ethics* 18 (1): 36–54.
Mazor, J. 2013. The Child's Interests and the Case for the Permissibility of Male Infant Circumcision. *Journal of Medical Ethics* 39 (7): 421–428. https://doi.org/10.1136/medethics-2013-101318.
Mazor, J. 2019. On the Strength of Children's Right to Bodily Integrity: The Case of Circumcision. *Journal of Applied Philosophy* 36 (1): 1–16. https://doi.org/10.1111/japp.12275.
Mazor, J. 2024. On Hypothetical Consent, Regret, and the Capacity for Autonomy: A Response to Pugh's Conceptual Analysis of the Child's Right to Bodily Integrity. *Clinical Ethics*, online ahead of print at https://doi.org/10.1177/14777509241260l7.
Merkel, R., and H. Putzke. 2013. After Cologne: Male Circumcision and the Law. Parental Right, Religious Liberty or Criminal Assault? *Journal of Medical Ethics* 39 (7): 444–449.
Möller, K. 2017. Ritual Male Circumcision and Parental Authority. *Jurisprudence* 8 (3): 461–479.
Munzer, S. R. 2018. Examining Nontherapeutic Circumcision. *Health Matrix* 28 (1): 1–78.
Muschialli, L., C. L. Allen, E. Boy-Mena, A. Malik, C. Pallitto, Å. Nihlén, and L. Gonsalves. 2024. Perspectives on Conducting "Sex-Normalising" Intersex Surgeries Conducted in Infancy: A Systematic Review. *PLOS Global Public Health* 4 (8): e0003568.
Newman, L. E., and E. N. Dorff. 2024. *Ethics at the Center: Jewish Theory and Practice for Living a Moral Life*. Philadelphia: Jewish Publication Society. https://muse.jhu.edu/pub/199/monograph/book/118055.
O'Neill, S., D. Bader, C. Kraus, I. Godin, J. Abdulcadir, and S. Alexander. 2020. Rethinking the anti-FGM Zero-Tolerance Policy: From Intellectual Concerns to Empirical Challenges. *Current Sexual Health Reports* 12:266–275.
Paatwala, S. 2018. *A Dawoodi Bohra Woman's Experience of Female Genital Cutting [Internet]*. Available from: https://www.youtube.com/watch?v=JZSpb9RVqFQ.
Porat, I. 2021. The Starting at Home Principle: On Ritual Animal Slaughter, Male Circumcision and Proportionality. *Oxford Journal of Legal Studies* 41 (1): 30–58.
Pugh, J. 2023. The Child's Right to Bodily Integrity and Autonomy: A Conceptual Analysis. *Clinical Ethics*, online ahead of print https://doi.org/10.1177/14777509231118881.
Rawls, J. 1997. The Idea of Public Reason Revisited. *The University of Chicago Law Review* 64 (3): 765–807.
Reis, E. 2021. Medicalization and the Mainstreaming of Circumcision in Mid-Twentieth-Century America. *American Jewish History* 105 (3): 321–342.
Remennick, L. 2022. Joining the Tribe: Adult Circumcision Among Immigrant Men in Israel and Its Traumatic Aftermath. *Culture, Health and Sexuality* 24 (5): 702–716.
Rizalar, S., E. T. Buyuk, and N. Yildirim. 2017. Children's Perspectives on the Medical and Cultural Aspects of Circumcision. *Iranian Journal of Pediatrics* 27 (2): 1–7.
Rogers, J. B. 2019. What Constitutes Mutilation? A coNcern with Anti-Female Genital Mutilation Laws in Australia and the Question of Natural Function. *Current Sexual Health Reports* 11:442–446.
Rosner, F. 1971. Autopsy in Jewish Law and the Israeli Autopsy Controversy. *Tradition: A Journal of Orthodox Jewish Thought* 11 (4): 43–63.
Sarajlic, E. 2014. Can Culture Justify Infant Circumcision? *Res Publica* 20 (4): 327–343.

Shweder, R. A. 2009. Shouting at the Hebrews: Imperial Liberalism vs. Liberal Pluralism and the Practice of Male Circumcision. *Law, Culture and The Humanities* 5 (2): 247–265. https://doi.org/10.1177/1743872109102491.

Shweder, R. 2017. The Goose and the Gander: The Genital Wars. In *Circumcision, Public Health, Genital Autonomy and Cultural Rights*, 141–159. Routledge.

Shweder, R. 2022. The Prosecution of Dawoodi Bohra Women: Some Reasonable Doubts. *Global Discourse* 12 (1): 9–27. https://doi.org/10.1332/204378921X16141809582432.

Svoboda, J. S., and R. S. Van Howe. 2013. Out of Step: Fatal Flaws in the Latest AAP Policy Report on Neonatal Circumcision. *Journal of Medical Ethics* 39 (7): 434–441.

Tao, S. 2024. *Uncertainty and Disquiet: My Non-Consensual Circumcision as a Trans Woman.* Autostraddle. Feb 20. https://www.autostraddle.com/uncertainty-and-disquiet-my-non-consensual-childhood-circumcision-as-a-trans-woman/.

Taylor, J. R., A. P. Lockwood, and A. J. Taylor. 1996. The Prepuce: Specialized Mucosa of the Penis and Its Loss to Circumcision. *British Journal of Urology* 77 (2): 291–295.

Townsend, K. G. 2023. Defending an Inclusive Right to Genital and Bodily Integrity for Children. *International Journal of Impotence Research* 35 (1): 27–30. https://www.nature.com/articles/s41443-021-00503-x.

Uberoi, M., J. Abdulcadir, D. A. Ohl, J. E. Santiago, G. K. Rana, and F. W. Anderson. 2023. Potentially Under-Recognized Late-Stage Physical and Psychosexual Complications of Non-Therapeutic Neonatal Penile Circumcision: A Qualitative and Quantitative Analysis of Self-Reports from an Online Community Forum. *International Journal of Impotence Research* 35 (3): 234–241.

Ungar-Sargon, E. 2018. *No Sacred Cows: A Conversation with Rabbi Asher Lopatin.*

Van Eekert, N., H. Barrett, S. Kimani, I. Hidayana, and E. Leye. 2024. Rethinking the Definition of Medicalized Female Genital Mutilation/Cutting. *Archives of Sexual Behavior* 53 (2): 441–453.

Van Howe, R. S. 2011. The American Academy of Pediatrics and Female Genital Cutting: When National Organizations are Guided by Personal Agendas. *Ethics and Medicine* 27 (3): 165–173.

Van Howe, R. S. 2013a. Infant Male Circumcision in the Public Square: Applying the Public Reason of John Rawls. *Global Discourse* 3 (2): 214–229.

Van Howe, R. S. 2013b. Infant Circumcision: The Last Stand for the Dead Dogma of Parental (Sovereignal) Rights. *Journal of Medical Ethics* 39 (7): 475–481.

Van Howe, R. S. 2018. Response to Vogelstein: How the 2012 AAP Task Force on Circumcision Went Wrong. *Bioethics* 32 (1): 77–80.

Van Howe, R. S., J. S. Svoboda, J. G. Dwyer, and C. P. Price. 1999. Involuntary Circumcision: The Legal Issues. *BJU International* 83 (S1): 63–73.

Viens, A. M. 2017. *The Right to Bodily Integrity*. Routledge.

WHO. 2008. *Male Circumcision: Global Trends and Determinants of Prevalence, Safety and Acceptability (No. UNAIDS/07.29 E/JC1320E)*. World Health Organization.

WHO. 2023. *Female Genital Mutilation*. Retrieved from: https://www.who.int/news-room/fact-sheets/detail/female-genital-mutilation.

Yavuz, F., and H. Kobya Bulut. 2024. The Effect of Circumcision Age on Anxiety and Self-Perception in Boys. *Comprehensive Child and Adolescent Nursing* 47 (3): 214–223.

Zakir, M. 2016. *WHO's Stance and the Criminalization of Female Circumcision: The Protection of or Violation of Human Rights?* Dawoodi Bohra Women's Association for Religious Freedom. https://dbwrf.org/uploads/cause_files/the-protection-of-or-violation-of-human-rightspdf_1517826222.pdf.

Ziemińska, R. 2020. The Epistemic Injustice Expressed in "Normalizing" Surgery on Children with Intersex Traits. *Diametros* 17 (66): 52–65.

Chapter 12
Contested Cuttings: The Ethics of Genital Cutting in Islam Through the Lens of Peer Disagreement and Secular Bioethics

Hossein Dabbagh

Abstract In recent years, there has been a resurgence in debates on the ethics of child genital cutting practices, both female and male, including within a Muslim context. Opponents of female genital cutting sometimes assert that the practice is not mentioned explicitly in the Qur'an as a way of implying that it does not have any religious standing within Islam. However, neither is male genital cutting mentioned explicitly in the Qur'an, and yet most people accept that it is a Muslim religious practice. Both practices, however, are mentioned in secondary sources of Islamic jurisprudence, with disagreement among religious authorities about the status or authenticity of some of these sources. This chapter considers the religious status of both female and male genital cutting practices within Islam and employs a philosophical argument from "peer disagreement" to ask whether either practice is necessary (i.e., religiously required) for a devout Muslim to endorse. To extend this analysis, the chapter incorporates secular bioethical perspectives, focusing particularly on autonomy, harm, and the tensions between cultural norms and individual rights, aiming to contribute to the ongoing dialogue on the coexistence of religious and secular viewpoints in bioethical debates.

An earlier version of this chapter appeared in the *Journal of Religion and Health*, an open-access publication from Springer Nature, under a Creative Commons Attribution 4.0 International License (Dabbagh, 2022). This updated version features a refined argument, introduces a new section, and is presented under a revised title.

H. Dabbagh (✉)
Associate Professor of Philosophy, Department of Philosophy, Northeastern University London, London, UK
e-mail: hossein.dabbagh@nulondon.ac.uk

K. R. Smith (ed.), *Bioethics and Religion*, The International Library of Bioethics 116, https://doi.org/10.1007/978-3-032-33538-8_12

12.1 Introduction

There are two different kinds of religious rituals within Islamic scriptural sources. The first kind are rituals mentioned explicitly in the Qur'an, and almost all Muslim scholars, both Shīʿite and Sunni,[1] agree about their importance and validity. However, the second kind are rituals about which Muslim scholars of similar knowledge, standing, or authority have peer disagreements about how to interpret them. In some cases, they may even have completely opposite judgments. In this chapter, I argue that rituals of the latter kind—that is, the ones that are subject to rational peer disagreement among similarly qualified scholars—can justifiably be seen, even within a purely religious worldview, as non-essential and contingent.[2] Consequently, a sincerely pious Muslim could be justified in declining to perform or endorse such a ritual. Furthermore, I will discuss how rational peer disagreement about the status of certain rituals within a Muslim religious context has significant philosophical and practical implications.

Among various rituals practised by Muslims, some such as Ṣalāt (daily prayer), Ṣawm (fasting), Ḥajj (pilgrimage), and Zakāt (alms tax) are generally non-controversial. However, practices such as circumcision (Khitān) and veiling (Hijāb) are two of the most controversial among Muslims and non-Muslims alike.[3] Although these are completely different practices, both circumcision and veiling raise a similar epistemic challenge[4]: Could a devout Muslim be religiously justified in declining to perform them, for example, on moral grounds? To be sure, the majority of Muslims believe that the Qur'an endorses ḥijāb for women as a veil that is supposed to cover the head and chest. However, a growing trend among Muslim scholars (either jurists or intellectuals)[5] argues that a head-covering ḥijāb is not *necessarily* religiously obligatory.[6] These intellectuals refer to certain interpretations of the Qur'an with an argument that ḥijāb essentially aims for modesty

[1] For more details on these two main branches of Islam and their philosophical and jurisprudential differences see: Gleave (2018) and Hallaq (1997, chs. 2-3).

[2] For more on the distinction between the essential and the accidental or contingent in Islam, see: Soroush (2009, Ch. 4) and Dabbagh (2023). Note that all rituals irrespective of their reference in the Qur'an are subject to rational peer disagreement among Muslims. Historically speaking, for example, not all Muslim communities have considered contingent what other Muslim communities may have considered essential. However, in this chapter, I will focus on those rituals that are not explicitly mentioned in the Qur'an.

[3] All rituals may be considered as practices, but not all practices may be considered as rituals. Circumcision may be considered as a ritual and a practice, but veiling is a practice rather than a ritual.

[4] There is also a difference between enforcing 'hijab' (which is textually anchored in the Qur'an) and children's circumcision, the latter is known to cause psychological damage to the future of children's well-being.

[5] Muslim intellectuals generally are not jurists. Both jurists (*ulemā*) and non-jurists are scholars though.

[6] See Mernissi (1992, Ch. 5). Shahrur (2009, Introduction and Ch. 5).

and a devoted religious woman can be a modest person even without wearing a ḥijāb.[7]

What about circumcision? Can a similar analysis be applied? While the majority of Muslims practice at least male circumcision, with a smaller proportion practising both female and male genital cutting in concert (see below), the question here is whether a well-informed, rational and perfectly devout Muslim could—using the hermeneutical resources of Islam itself—decline to perform or authorise the circumcision of a male child who is too young to consent for himself, as is widely accepted in the case of female children. I will answer in the affirmative.[8]

12.2 Circumcision: What Is It and Why Is It Contentious?

First, we must clarify what circumcision refers to. In English, the term 'circumcision' is most commonly used to describe male circumcision specifically, that is, the partial or total removal of the penile prepuce or foreskin. However, the Arabic word for circumcision —'khitān'— is gender-neutral and refers to the cutting of either the male or female prepuce (Box 12.1). In the Western world, any cutting of female genitalia that is not considered medically necessary[9] is usually termed 'female genital mutilation' (FGM), following the World Health Organization's definition (WHO 2008). However, this definition includes some forms of ritual female genital cutting (FGC) that are less physically substantial than the male circumcisions carried out within the same families, raising questions about sex-based categorisations (O'Neill et al. 2020). According to one interpretative tradition, girls within the practicing subsets of Muslims are considered equally 'worthy' of being circumcised as boys (albeit by means of a less intrusive procedure), thus marking a break with the older Jewish covenantal ritual, from which girls are, by contrast, excluded—arguably due to having a lower status than males within classical rabbinical Judaism (Shweder 2022; Cohen 1997). In the case of Islam, the forms of FGC in question include the so-called ritual "nicking, pricking, or partial removal of the clitoral prepuce or hood"—the most common forms of FGC in some parts of South and Southeast Asia, where they are carried out, alongside male circumcision, for religious reasons within some sects of Islam (Earp 2022b; Earp et al. 2021; Dawson et al. 2020; Duivenbode and Padela 2019; Bootwala 2019; Rashid et al. 2020).

---

[7] Ahmad Qabil, as a prominent Shīa scholar, thinks so. See his e-book *Shari'at-i 'aqlani* [Rational Religion] on his website: http://www.ghabel.net/shariat/. See also Ridgeon (2020) and Rezaei (2020).

[8] My argument here is philosophical. It does not necessarily answer psychological and sociological questions, such as why Muslims still practice circumcision if it is not religiously necessary.

[9] See BCBI (2019) as a recent international consensus statement.

Box 12.1 A brief overview of the human prepuce: male, female, intersex. Adapted with permission from Myers and Earp (2020) and Earp (2022)

The genital prepuce is a shared anatomical feature of both male and female members of all human and non-human primate species (Cold and Taylor 1999). In humans, the penile and clitoral prepuces are undifferentiated in early fetal development, emerging from an ambisexual genital tubercle that is capable either of penile or clitoral development regardless of genotype (Baskin et al. 2018). Even at birth—and thereafter—the clitoral and penile prepuces may remain effectively indistinguishable in people with certain intersex traits or differences of sex development (Hodson et al. 2019; Fahmy 2015; BCBI 2025; Pippi Salle et al. 2007). The prepuce is an integrated feature of the external genitalia, having evolved to function in concert with other genital structures; for example, it forms the anatomical covering of the glans penis or clitoris, thereby internalizing each and 'decreasing external irritation and contamination' (Cold and Taylor 1999, p. 34). In the case of the penile prepuce, an additional function—alongside its biomechanical role in sexual intercourse (Purpura et al. 2018)—is to protect the urinary opening from abrasion, as this runs through the penile, but not the clitoral glans (Fahmy 2020). The penile prepuce has a mean reported surface area of between 30 and 50 square centimeters in adults (Kigozi et al. 2009; Werker et al. 1998) and is the most sensitive part of the penis, both to light touch stimulation and sensations of warmth (Bossio et al. 2016; Sorrells et al. 2007). The clitoral prepuce, while smaller in absolute terms, is continuous with the sexually-sensitive labia minora; it is also an important sensory platform in its own right, and one through which the clitoral glans can be stimulated without direct contact (which can be unpleasant or even painful) (O'Connell et al. 2008). In both sexes, the human prepuce is 'a specialized, junctional mucocutaneous tissue which marks the boundary between mucosa and skin [similar to] the eyelids, labia minora, anus and lips... The unique innervation of the prepuce establishes its function as an erogenous tissue' (Cold and Taylor 1999, p. 34).

In deciding what terminology to employ in this context, it is important to note that ritual male genital cutting (MGC), in contrast to ritual FGC, has not been defined as 'mutilation' by any Western organization even in its most dangerous forms (O'Neill et al. 2020). For example, ritual male circumcision as practiced by the Xhosa of South Africa frequently leads to scarring, perceived disfigurement, and carries a high rate of penile amputation and death, often due to sepsis or suicide (Earp et al. 2017; Douglas and Nyembezi 2015; van der Merwe 2020).[10] The reason

[10] These and other examples of relatively severe forms of male genital cutting are summarized in Shahvisi and Earp (2019). For further inconsistencies in the WHO policies on child genital cutting practices, see Earp and Johnsdotter (2021).

why no form of medically unnecessary MGC, but all forms of medically unnecessary FGC, are considered to be 'mutilating' by the WHO, irrespective of the actual extent of cutting in either case or any associated physical-functional implications, remains unclear.[11] However, some scholars argue that such disparate labelling may reflect a Western evaluative bias,[12] given that MGC, but not FGC, is familiar to Western culture (Darby 2016; Njambi 2004; Oba 2008; Tangwa 2004). Historically, this would have been due to contact with Judaism (Glick 2005), but the driving force today seems to be medicalized newborn male circumcision as it is practiced in the United States and to a lesser extent in Canada (Earp 2016; Gollaher 2000; Darby 2013). As Toubia (1999, p. 5) notes:

> A major difference between male and female circumcision is that the female procedure is primarily carried out in Africa [among other regions of the Global South], which is currently the least dominant culture in the world. The male procedure is also common in the same countries, but it is also common in the United States, which is currently the most dominant culture in the world through its far-reaching media machine. This historical situation has made it easier to vilify and condemn what is common in Africa and sanctify what is popular in America.[13]

Within a Muslim religious context, both male and female circumcision are controversial as neither is ordained nor even mentioned *explicitly* in the Qur'an. However, in addition to the Qur'an, there are other sources of Islamic legislation, such as prophetic Hadiths (i.e., sayings and deeds attributed to the Prophet). The Qur'an itself does not contain all relevant rulings. Many issues in Islamic Sharīʿa law rely not on the Qur'an but on the Hadiths. For example, although the prayers, subject to interpretation, are considered by many as one of the "essential" pillars of Islam, the number of prayers, their parts, and how to pray are not mentioned in the Qur'an. Circumcision is an analogous issue in the sense that the primary source of its interpretation is Hadith.

There are two main schools of thought regarding male circumcision in Islamic jurisprudence. Some hold that it is a religiously obligatory practice (the majority position), and others hold that it is merely preferred or recommended (argued by a few jurists). For example, Mulla Muhsin al-Fayḍ al-Kāshānī, in his *al-Maḥajjat al-bayḍā'*, references Al-Ghazali's *Iḥyā' 'ulūm al-dīn* to report various Hadiths on male circumcision. Some Hadiths classify circumcision as *wājib* (obligatory), a religious duty commanded by God, so the one who does it will deserve a reward, and the one who fails to do it will be punished in the afterlife. Others consider it

[11] I will refer to both sets of practices as "genital cutting" or as "circumcision" where appropriate, preferring to leave the decision about whether one's own altered genitalia should be described as mutilated to each affected individual (see Earp 2019).

[12] See Johnson (2010).

[13] Another potential difference is that MGC, but not FGC, has become culturally associated with health benefits in Western contexts, including a reduction in the risk of acquiring certain infections (although the existence, likelihood, and/or magnitude of these benefits is controversial in the medical literature). For an alternative view, see Earp et al. (2021).

mustahabb (recommended), meaning that it is not essential to be done though its fulfilment is rewarded, and its negligence is not be punishable (1960, vol. 1, 524–526).

It is also reported from the Prophet in *Ṣaḥīḥ al-Bukhārī* (5891) that male circumcision, along with shaving the pubic hair, is among those activities recommended for a Muslim to do for hygiene and purification. Similarly, in answering the question of whether male circumcision is required of parents to do or authorize for their children, Ayatollah Sistani, a prominent Shīa Marji' Taqlīd (religious authority), replies that it is recommended and not obligatory.[14] From this perspective, one could argue that male circumcision *can* be deferred until the boy reaches puberty and can decide for himself.[15]

By contrast, in Islamic jurisprudence, opinions endorsing female circumcision are typically regarded as less authoritative than those endorsing male circumcision. This is partly because while male circumcision has been mentioned in many prophetic Hadiths, the Hadiths concerning female circumcision are relatively few in number and can more readily be judged as weak or dubious according to rules established by scholars of the Hadiths.[16]

Another argument for distinguishing male and female circumcision in terms of their respective statuses in Islam, considered globally, lies in their prevalence. Female circumcision is not a common practice in most Muslim-majority countries, whereas male circumcision is ubiquitous within the same countries. This leads to the common argument that "female circumcision is not really a 'religious' practice" (i.e., inherent to, or necessary within, the religion; otherwise, it would be more widespread) "but is rather 'merely' a cultural practice" that has only incidentally become associated with Islam in certain contexts (Earp 2021).[17] In support of this view, it is sometimes mentioned that female genital cutting existed in Africa prior to the advent of Islam and was incorporated into Muslim ritual practice in some, rather than all, places where Muslims have large established populations.

A potential problem with this argument is that male genital cutting also existed in Africa prior to the advent of Islam—and indeed Judaism—and was eventually adopted by various Semitic tribes, including the ancestors of both Muslims and Jews (Cohen 1997). Moreover, there may be alternative explanations[18] for the

---

[14] See his website: https://www.sistani.org/persian/qa/0896/

[15] See Dabbagh (2017).

[16] Gomaa (2013).

[17] See Bewley and Momoh (2010). Also see: Odukogbe et al. (2017).

[18] In brief, one explanation holds that males have had much greater social, political, and legal status than females throughout most of the respective histories of Judaism and Islam. Since circumcision usually is considered a status-elevating rite within practicing groups (e.g., a sign of a divine covenant within Judaism), women and girls have more typically been left out. With respect to Jewish practice, for example, as Cohen (1997) notes: "Of all the rituals from which women are excluded by rabbinic culture, the exclusion from circumcision is at once the most obvious and the most problematic" (p. 560). This suggests that it is the exclusion of females from ritual circumcision that "reflects their lower status" in certain contexts (Earp 2022b). Nevertheless, in the subset of Muslim

near-universality of male, but not female, circumcision within these faith communities that do not amount to a dichotomous distinction between 'religious' (male) and 'merely cultural' (female) genital cutting in situations where Muslims practice both together. Indeed, the very distinction between 'religious' and 'cultural' as descriptions of certain practices has been questioned by anthropologists and philosophers. For example, as Brusa and Barilan (2009) argue:

> In the context of circumcision, we find the moral dimension of the distinction between the 'religious' and the 'cultural' rather tenuous. Research on the historical development of circumcision demonstrates very intricate links bridging religion, institutions of social power and metamorphoses in meaning and practice over time and space (p. 471).

Nevertheless, many Muslim scholars, including jurists and intellectuals, believe that female circumcision is not a religious obligation for devout Muslims. Ayatollah Sistani, for instance, in a previously mentioned fatwā, states that female circumcision is *haram* (forbidden) if it causes harm. Whether it is harmful or harmful enough to fall within this proscription is a matter of ongoing debate (see, e.g., Rogers 2016). Similarly, Sheikh Muhammad Sayyid Tantawy of Al-Azhar contends that female circumcision is an un-Islamic ritual, unrelated to the religion.[19]

In line with this perspective, I will assume that a devout Muslim could be justified in declining to perform or authorize such a procedure for their daughter. In the case of male circumcision, however, such an argument seems harder to make. This is because the majority opinion among Muslim jurists and intellectuals with respect to male circumcision is the very opposite of what I have just said about female circumcision: they regard it as religiously required. However, I propose that an argument can be made against this requirement, employing the logic of peer disagreement, which I will explain in the following sections.[20]

12.3 Grappling with Peer Disagreement

As noted earlier, the main reason for differing opinions on circumcision is that circumcision, both female and male, is not mentioned explicitly in the Qur'an. When a matter (e.g. circumcision) is not explicitly covered in the Qur'an, Muslim jurists derive their rulings (ahkam) from other sources, employing the principles and methodology of Islamic jurisprudence (*usul al-fiqh*). These sources include the Sunnah (Hadith) which is what the Prophet—and *Imams* in the Shīʿite school—reportedly said, did or agreed to; *ijma*, the consensus of Islamic scholars; and '*aql*, which is the power of reason or the rational mind (although 'aql is not generally recognised a

communities where they are, in fact, included in the institution of circumcision alongside males, it does not follow that the practice ceases to be 'religious' in nature only in the case of females.

[19] See http://news.bbc.co.uk/1/hi/42914.stm. See also Kassamali (2005).

[20] See Najafi (1981, Vol. 31, pp. 260-263 and 257) and Alahmad and Dekkers (2012).

source of Sunni *usul al-fiqh*). The absence of circumcision in the Qur'an makes it liable to different interpretations and opinions among scholars.

On one interpretation, based on Shīʿite and Sunni jurisprudential texts, male circumcision before the age of seven is not obligatory in Islam; it is *merely* preferred or recommended (*mustahabb* or *mandub*), or considered a kind of custom, according to the *ulemā* (Muslim jurists). In addition, men converting to Islam *could* be circumcised if they wished; hence there is no strict requirement to be circumcised.[21]

However, many *fuqahā* (jurists) agree that male circumcision is obligatory for those who want to become an imam in performing prayer, i.e. the one who leads the congregation in ṣalāt.[22] It is also agreed that circumcision (primarily for men) is obligatory during Hajj.[23] In effect, as one of the Islamic rituals of pilgrimage, Muslims who want to go to Mecca for Ṭawāf during Hajj are required to be circumcised beforehand. However, since Hajj is not *wājib* (obligatory) for those unable to afford it, and becoming an *imam* is voluntary, one might argue that circumcision is not *always* obligatory for all Muslims.

Sunni and Shīʿite jurists have different opinions on whether male circumcision is obligatory or an act of Sunnah—i.e., not obligatory but highly recommended. Among Sunni schools, while Shafiites generally view male circumcision as obligatory, Hanbalites, Ḥanīfis and Mālikis consider male circumcision potentially obligatory. Notably, Imam Abū Ḥanīfa and Imam Mālik categorized male circumcision as a confirmed Sunnah (*Sunnah Mu'akkadah*)—which is highly recommended but not obligatory (al-Ḥaṭṭāb 1995; al-Humām 1999; Alahmad and Dekkers 2012). Some Shīʿite jurists, but not all, regard male circumcision before the age of seven as non-obligatory but preferred or recommended (*mustahabb* or *mandub*).[24]

Given this understanding, it seems plausible to argue that since the necessity of circumcision in Islamic *Sharīʿa* law both for male and female Muslims has been challenged by a number of Shīʿite and Sunni *ulemā*, it may be the case that being circumcised, or authorizing circumcision for one's children, is not a necessary part of being a Muslim. In other words, one can be a devout Muslim—in good standing, so to speak—and yet not practice the circumcision ritual simultaneously. But this raises an important philosophical question: why, exactly, does disagreement among *ulemā* provide us a reason to conclude that circumcision might be unnecessary?

---

[21] See Khomeini (1987, Vol. 2, p. 310). See also: Najafi (1981, Vol. 31, pp. 262-3) and Ameli (1994, Vol. 24, p. 271).

[22] See Najafi (1981, Vol. 13, pp. 384-385).

[23] See Najafi (1981, Vol. 19, p. 274). However, there are some alternative views on whether circumcision is required for Hajj. See, e.g., Ameli (1991, Vol. 8, p. 118).

[24] See Najafi (1981, Vol. 41, 672) and Tusi (1973, Vol. 8, 67).

12.4 Peer Disagreement Regarding Harm Being Divinely Ordained[25]

What is the difference between rituals that are mentioned explicitly in the Qur'an and practices that are not mentioned? Does this distinction hold any significant philosophical or theological implications? This question becomes particularly relevant in recent discussions within social epistemology concerning "peer disagreement" (*ikhtilāf al-aqrān*).[26] In Islam, rational peer disagreements are especially likely to arise when we have rituals and practices that are not explicitly detailed in the Qur'an.[27] The peer disagreements will be especially intense if the notion of harm is involved. For example, in the debate on circumcision, there is a possibility that causing harm to a child is, in fact, permissible if divine ordination can be ascertained, while others would argue that a practice cannot be imposed on children if it is harmful on balance or overall. Let us first explain what peer disagreement is.

Consider this scenario: Suppose person A believes proposition p while person B believes not-p. Suppose further that they are referring to religious texts as evidence supporting their respective claims and that they have equal reasoning skills. Given that p and not-p are contradictory (based on classical logic's law of non-contradiction), either A or B must be wrong. However, if both A and B are rational based on their *prima facie* justification, with equal reasoning skills (by stipulation), they should respect one another as reasoners. What should be done in the face of such rational peer disagreement? How should it be resolved?

Given that A and B have the same primary textual evidence, one approach is that they should proceed to revise their initial beliefs in light of other sources such as ethics, philosophy, science, etc. A's and B's initial views about the proposition believed (p) carry equal weight, suggesting that, upon learning about the other's contradictory belief, they should become less confident in their own opinions. However, both A and B are rationally required to modify their views after a more comprehensive moral, philosophical, and scientific analysis. In this context, they might engage in a process known as *reflective equilibrium* (described below). This method helps in reaching a more sound and justifiable conclusion (Dabbagh et al. 2023). If there are still disagreements after further assessments, both parties then have "equal weight" views. It is important to recognize that not *every* form of disagreement necessitates a rational change in our beliefs, but reflective equilibrium provides a structured way to evaluate when such change is warranted.

The reflective equilibrium method seeks to achieve coherence, normatively, between our considered beliefs and judgments about particular cases and the

---

[25] Parts of my argument in this and the following section were previously discussed in an interview with Helen De Cruz for the Prosblogion, a blog focused on the philosophy of religion.

[26] Christensen (2007) and Christensen (2013).

[27] There are also some peer disagreements about rituals and practices that are mentioned in the Qur'an. For instance, as noted above, although ḥijāb is mentioned in the Qur'an, there are different views as to whether covering the head is necessary.

general theoretical principles or rules that explain or justify such judgments. We aim to revise our particular beliefs and judgments, and general principles through reflective equilibrium to reach a plausible and rational coherence among them. As Gaut describes:

> reflective equilibrium is not a theory, but a method, of justification; it tells one how to justify one's moral beliefs, by attempting to render consistent one's moral principles with one's judgments about particular cases (Gaut 2002, 139–140).

The process of reflective equilibrium is not solely for modifying our prior beliefs. Rather, we can also add new normative beliefs to our theoretical framework. As Scanlon (2002) notes, in practical contexts, reflective equilibrium can guide us in reaching a conclusion about what we *should* do when we are unsure about the right course of action.

In the pursuit of reflective equilibrium, individuals like A and B should seek coherence between their beliefs across various levels of generality. In effect, their very general beliefs need to align with their less-general beliefs and so on, down to their particular judgements. Importantly, they can achieve justification through reflective equilibrium even if they do not have conclusive evidence that they are correct. This entails that the pursuit of reflective equilibrium sometimes reveals new intuitions or judgments we had not previously made. For example, person A might have had a belief about the necessity of performing circumcision in every situation, but then is brought to see that there may not be such a strict duty, insofar as he maintains his firm moral judgments about broader principles regarding unnecessary harming, or in reasoning by analogy to a wider range of cases, and wishes to have those broader principles make sense. Similarly, person B might not initially have seen a moral duty not to inflict unnecessary harm. Still, she comes to see there must be such a duty for her firm moral intuitions about various particular cases to be correct. And this, in turn, might count against the belief that circumcision is *always* a duty even if it is divinely ordained. Through reflective equilibrium, both A and B refine their beliefs, achieving a more nuanced understanding that aligns with their broader moral principles.

Both A and B are grappling with the issue of divine command and the possibility of harm being divinely ordained. They are dealing with an example of inflicting harm (in the sense of a bodily injury, accompanied by pain, effectuated through cutting as in circumcision) and asking whether God's command—if that's what it is—can be justified. However, they might ask themselves what should be done when God tells us to cause harm, and His command does not seem morally to make sense (e.g., to injure and cause pain to a sensitive part of the body when it is not medically necessary to do so). Although they might have different strategies to check and balance their religious beliefs or interpretations with their moral judgments, in any case, they need to seek reflective equilibrium.

For instance, suppose A believes in strictly adhering to the general principle of *always* obeying the commands of Sharīʿa laws because he is convinced that God fully endorses these Sharīʿa laws and that obeying the commands of Sharīʿa laws is inherently morally good. On the other hand, B contends that obeying divine command is not *always* morally obligatory, especially if we have a strong moral reason, *ceteris paribus*, against a practice that inflicts unnecessary or unjustifiable harm. These views may come into conflict. If they do, A and B should seek a reflective equilibrium between their beliefs. A will have several choices. He can revise his particular belief to be better justified rationally (for example, discarding a particular Sharīʿa law if there are strong moral reasons against it), or modify his general principle (for example, only obeying the Qur'an or adopting a different interpretation of circumcision). Some Shīʿite scholars believe that some judgments such as jihād and Friday ṣalāt can be suspended until Imam Mahdi, the prophesied redeemer of Islam who is in occultation, arrives.[28] Person A might conclude in favour of suspending circumcision following this line of argument. B, meanwhile, might need to reconsider his stance on excessive rationalism, if at all. In situations where A's and B's beliefs are mutually exclusive, then both should diminish their confidence in their own positions and/or explore alternative interpretations to find a middle ground.

12.5 Sociocultural Understanding of Harm

Let us now turn to the question of 'harm' in the context of circumcision. One might wonder how harm should be interpreted in this context when harm is, allegedly, being divinely ordained? What are the implications for male or female circumcision? There are many Prophetic sayings about which scholars have disagreements concerning aetiology, authenticity and reliability. For example, Abū Ḥanīfa, the eponymous founder of the Sunni Ḥanafi School, believed that there are only a few Hadiths that we can historically trust (Syamsuddin 2001). One such widely cited Hadith, which almost all Islamic scholars refer to and on which many *Sharīʿa* laws are based, is the "principle of harm": let there be no harm nor reciprocating harm (*lā darar wa lā dirār*).

Subjectively speaking, many circumcised Muslims, whether male or female, do not seem to regard themselves as net harmed or harmed on balance by virtue of their circumcision. Some may not regard themselves as harmed, even in a purely physical sense. Others may concede that, to some extent, their body has been harmed while still believing that psychosocial or religious benefits outweigh those harms so that they are not harmed in the aggregate or overall. In Islamic *usul al-fiqh*, it is generally held that cutting any part of the body in itself is *prima facie* forbidden due to its harmful nature. However, it becomes permissible if there is a utility (*maslaha*) that outweighs the harm. So, cutting any part of the body is *only* permissible if its expected utility outweighs its harm. To understand this phenomenon, one needs to

[28] For a similar discussion, see: Alishahi Tabriz et al. (2016). See also Ghassemi (2009) and Akbar (2018).

have a nuanced account of harm and to understand the sociocultural and epistemic context in which circumcision is practiced.[29]

According to philosopher Scott Campbell (2006, pp. 226–227):

> one of the components of risk is harm. This is the level of badness or loss associated with the occurrence of x. Harm does not just include physical injury but any sort of circumstance that P would prefer not to be the case. If there is nothing bad about x at all then P is not at risk from x. This entails that our attitudes and preferences partly determine risk, because our attitudes and preferences determine what counts as a harm.

Here 'P' refers to a person and 'x' to an action or occurrence. In Campbell's view, once we understand our subjective attitudes and preferences, we will be able to identify the objective nature of certain harms. Consequently, Campbell argues that it is often impossible in many cases to determine for sure whether someone is at risk of harm or has actually been harmed until we know their relevant (considered) attitudes and preferences. With respect to circumcision, one's attitudes and preferences will likely be informed by what is culturally normative in their environment; but attitudes and preferences vary both within and between cultures, and they are subject to changing over time, especially as one acquires new information or learns different perspectives. For instance, if one grows up in a culture where the prepuce, whether male or female, is widely believed to be a "useless flap of skin," one may intuitively assign a value of zero to this tissue, such that its being damaged or removed will not seem a harm. Yet, exposure to different cultural assumptions and practices about genitalia—through travel, reading, or online exploration—can lead people to re-evaluate these beliefs and their own genital status, potentially seeing the predominant practice in their culture as harmful or problematic (Earp and Darby 2017, p. 25; see also Barutçu 2023).

For example, a man might discover that the penile foreskin, rather than being a "useless flap of skin," may be "the most sensitive part of the penis to light-touch sensation" (Bossio et al. 2016; Sorrells et al. 2007; Earp 2022), or that its manipulation in sexual contexts affords particular subjective sensations that many men with foreskins report finding valuable (Ball 2006). If such a man reached the conclusion that, in fact, he had been harmed by virtue of having his foreskin removed, this would not be obviously unreasonable.[30] Analogously, consider this account from a woman in the Dawoodi Bohra community who, in adulthood, came to reflect on her childhood circumcision and see it in a new light (quoted in Taher 2017):

> … I feel robbed and cheated of my sexuality, and feelings of inadequacy and incompleteness remain with me till today, even at the age of 61.

These examples illustrate how changes in information, cultural exposure, and personal reflections can lead to a reevaluation of what constitutes harm in the context of circumcision.

[29] For more on this, see: Sachedina (2009, Introduction).

[30] Concerning the potential damage that can be caused in male circumcision, please see this personal story about a botched circumcision in the *New Yorker*: https://www.newyorker.com/magazine/2021/10/11/a-botched-circumcision-and-its-aftermath

Insofar as it is reasonable for a Muslim person, whether male or female, to regard themselves as having been harmed by virtue of having had their genital prepuce cut for ritualistic reasons as a child, the "principle of harm" Hadith might seem to weigh against the practice of non-voluntary and non-therapeutic circumcision. However, consider someone who believes that the potential benefits of male circumcision, such as reducing the transmission of sexually transmitted infections, including HIV (as endorsed by the WHO for adult, voluntary circumcision), outweigh the harms associated with performing circumcision. In such a case, the individual is entitled to act on his justified belief as long as he is adequately informed and makes this decision *autonomously*.

12.6 Exploring the Intersection: Islamic and Secular Bioethics

Examining the nuanced landscape of Islamic jurisprudence and ethical reasoning alongside secular bioethics reveals insightful perspectives, especially concerning autonomy, harm, and the balance between cultural and individual rights.

Secular bioethics, which emphasizes individual autonomy and the avoidance of harm, can offer a practical lens through which to view the practice of circumcision. Autonomy, a cornerstone of secular bioethics, asserts the importance of an individual's right to make decisions about their own body, free from coercion. This principle raises significant questions about the ethics of performing irreversible bodily modifications, such as circumcision, on individuals unable to consent such as circumcision of minors (Beauchamp and Childress 2013, Ch. 4). The debate here intersects with Islamic perspectives on bodily integrity and parental rights, creating a complex dialogue between religious doctrine and secular principles of individual rights. Whitin many Islamic communities, circumcision functions as a rite of passage rather than a purely bodily practice, imbued with religious and cultural significance. From the perspective of secular bioethics, however, performing circumcision requires careful consideration of how autonomy, informed consent, parental decision-making, and cultural tradition should be balanced, particularly when it comes to irreversible procedures performed on those who are not yet capable of making such decisions for themselves (e.g., children).

Further, the secular bioethical principle of non-maleficence, or the obligation to avoid causing harm, invites a critical examination of the physical and psychological impacts of circumcision. While some argue that circumcision carries potential health benefits, secular bioethics would demand serious scrutiny of these claims, weighing them against the potential physical and psychological harms of the procedure and the violation of bodily integrity (Macklin 2003). Balancing these considerations requires a nuanced understanding of the harm that incorporates both the immediate physical outcomes and the longer-term psychological and social effects.

This scrutiny aligns with Islamic ethical considerations about harm but through a lens focused more on individual well-being than communal or religious norms.

The dialogue between cultural and individual rights is another critical intersection between secular bioethics and the religious discourse on circumcision. Cultures and religions often prescribe practices seen as beneficial or necessary for collective well-being and identity. However, secular bioethics challenges us to consider the individual's rights in the face of cultural norms, and the resulting tension between cultural tradition and individual rights requires careful consideration of how informed consent is obtained and respected, particularly when these practices have irreversible physical or psychological effects. Secular bioethics often prioritizes individual rights, which can conflict with cultural practices like circumcision that are embedded in religious and communal identities. The ethical deliberation thus expands to encompass the extent to which cultural practices should be upheld when they conflict with the individual's rights to bodily integrity (Gillon 2015).

In integrating these secular ethical considerations into the discussion of circumcision within the Muslim context, this chapter aims to bridge the gap between religious doctrine and secular bioethics. Incorporating these secular perspectives into the discussion on circumcision in Islam adds a multidimensional aspect to the ethical analysis, creating a more holistic understanding of religious and secular ethical considerations. By examining the practice through both lenses, we gain a more comprehensive understanding of the ethical complexities involved. This dual perspective not only enriches the ethical discourse but also aligns with the broader exploration of the coexistence of religious and secular perspectives. It underscores the necessity of engaging in ongoing dialogue and reflection on the ethical dimensions of circumcision, informed by a comprehensive view that respects both religious traditions and the principles of secular bioethics.

12.7 Conclusion

Peer disagreements are likely to arise when we have rituals that are *not* explicitly mentioned or prescribed in the Qur'an. In this chapter, I have argued that (1) such rituals, which are subject to rational peer disagreements among Muslim jurists/scholars, can plausibly be seen as non-obligatory, non-essential and contingent, (2) male circumcision is not explicitly prescribed in the Qur'an, and it is subject to rational peer disagreement, (3) therefore, male circumcision is non-obligatory, non-essential and contingent. Specifically, I have demonstrated that rational peer disagreement about the status of certain rituals within a Muslim religious context carries important philosophical and practical implications. Those religious rituals subject to rational peer disagreement among qualified scholars can plausibly be seen as non-obligatory and non-essential rituals to practice. The mere fact of having a rational peer disagreement can make us, at least, suspend extra-Qur'anic rituals. Thus, a devout Muslim could be rationally justified in declining to perform such rituals. The presence of rational peer disagreement across scholars can be a reason to reconsider the necessity of practicing such rituals. However, this does not entail *every* form of disagreement provides an acceptable reason to decline to act.

Following my argument in this chapter, we can conclude that religious practice in connection with circumcision (primarily male) among the larger Muslim community is not authoritative since the Qur'an does not prescribe it as a necessary practice. This makes it plausible to maintain philosophically (or rationally) that circumcision is not obligatory in one's claim to be a member of the Muslim community. Additionally, based on differences of opinion among Muslim jurists in regard to interpreting textual sources like the Sunnah, in the context of juridical methodology that applies the principle of avoiding harm and promoting the public good, the practice can be regarded as unnecessary or at best recommended (mandub, mustahabb) like a number of cultural practices in Muslim societies.

As far as circumcision is considered, my argument implies that it is not morally and religiously *necessary* to perform circumcision in Islam, given the lack of explicit mention in the Qur'an and the existence of rational peer disagreements among Muslim scholars about whether circumcision must be performed or not. Sunni and Shīʿite jurists and intellectuals have varied opinions on whether male or female circumcision is obligatory or Sunnah acts. So, it seems plausible to conclude that the necessity of circumcision in Islamic *Sharīʿa* law both for male and female Muslims has been challenged by a number of Shīʿite and Sunni *ulemā*, and that circumcision is not a *necessary* part of being a Muslim.

Acknowledgement I would like to express my gratitude to Brian D. Earp for his thoughtful and critical comments on this chapter's initial draft.

References

Akbar, A. 2018. The Classical Islamic Laws of Apostasy in the Present Context. *International Journal of Humanities and Social Sciences* 12 (6): 666–671.

Alahmad, G., and W. Dekkers. 2012. Bodily Integrity and Male Circumcision: An Islamic Perspective. *The Journal of IMA* 44 (1): 7903.

al-Ḥaṭṭāb, Muḥammad. 1995. *Mawāhib al-Jalīl li sharḥ Mukhtaṣar Khalīl*. Beirut: Dār al-Kutub al-ʻIlmiyya.

al-Humām, Muḥammad Ibn ʻAbd al-Wāḥid ibn. 1999. *Sharḥ Fatḥ al-Qadīr*. Beirut: Dār al-Kutub al-ʻIlmiyya.

Alishahi Tabriz, A., H. Dabbagh, and H. G. Koenig. 2016. Medical Ethics in Qiṣāṣ (Eye-for-an-Eye) Punishment: An Islamic View; an Examination of Acid Throwing. *Journal of Religion and Health* 55 (4): 1426–1432.

Ameli, M. 1991. *Madarek Al-Ahkam fi Sharh Sharaye Al-Islam*. Mashhad: Aaal Al-Bait Publisher.

Ameli, H. 1994. *Wasā'il al-Shīʿa fi Tahsil Masael Al-Sharia*. Ahl Al-Bait Qom Publisher.

Ball, P. J. 2006. A Survey of Subjective Foreskin Sensation in 600 Intact Men. In *Bodily Integrity and the Politics of Circumcision*, 177–188. Dordrecht: Springer.

Barutçu, A. 2023. 'Wounded Religious Masculinities': Muslim Men's Opposition Against Male Circumcision in Turkey. *Turkish Studies*, 24(2), 379–399. https://doi.org/10.1080/14683849.2022.2103408.

Baskin, L. S., J. Shen, A. W. Sinclair, M. Cao, X. Liu, G. Y. Y. Liu, D. Isaacson, M. Overland, Y. Li, and G. R. Cunha. 2018. Development of the Human Penis and Clitoris. *Differentiation* 103 (1): 74–85.

BCBI: Brussels Collaboration on Bodily Integrity. 2019. Medically Unnecessary Genital Cutting and the Rights of the Child: Moving Toward Consensus. *American Journal of Bioethics* 19 (10): 17–28.

Brussels Collaboration on Bodily Integrity. 2025. Genital Modifications in Prepubescent Minors: When May Clinicians Ethically Proceed? *The American Journal of Bioethics, 25(7), 53–102.* https://doi.org/10.1080/15265161.2024.2353823/ppbr//p

Beauchamp, T. L., and J. F. Childress. 2013. *Principles of Biomedical Ethics*. 7th ed. Oxford University Press.

Bewley, Creighton S., and C. Momoh. 2010. Female Genital Mutilation. *BMJ* 340:c2728.

Bootwala, Y. 2019. A Review of Female Genital Cutting in the Dawoodi Bohra Community: Part 3—The Historical, Anthropological and Religious Underpinnings of FGC in the Dawoodi Bohras. *Current Sexual Health Reports* 11 (3): 228–235.

Bossio, J. A., C. F. Pukall, and S. S. Steele. 2016. Examining Penile Sensitivity in Neonatally Circumcised and Intact Men Using Quantitative Sensory Testing. *The Journal of Urology* 195 (6): 1848–1853.

Brusa, M., and Y. M. Barilan. 2009. Cultural Circumcision in Eu Public Hospitals-An Ethical Discussion. *Bioethics* 23 (8): 470–482.

Campbell, S. 2006. Risk and the Subjectivity of Preference. *Journal of Risk Research* 9 (03): 225–242.

Christensen, D. 2007. Epistemology of Disagreement: The Good News. *Philosophical Review* 116 (2): 187–217.

Christensen, D. 2013. Epistemic Modesty Defended. In *The Epistemology of Disagreement*, edited by David Christensen and Jennifer Lackey, 76–97. Oxford: Oxford University Press.

Cohen, S. 1997. Why Aren't Jewish Women Circumcised? *Gender and History* 9 (3): 560–578.

Cold, C. J., and J. R. Taylor. 1999. The Prepuce. *BJU international* 83 (S1): 34–44.

Dabbagh, H. 2017. The Ethics of Non-Therapeutic Male Circumcision Under Islamic Law: A Lesson For Educational Prosperity In Muslim Communities. *TARBIYA: Journal of Education in Muslim Society* 4 (2): 216–223.

Dabbagh, H. 2022. Is Circumcision "Necessary" in Islam? A Philosophical Argument Based on Peer Disagreement. *J Relig Health* 61, 4871–4886. https://doi.org/10.1007/s10943-022-01635-0

Dabbagh, H. 2023. Epistemology of religion and phenomenology of revelation in post-revolutionary Iran: The case ofAbdolkarim Soroush. *Philosophy & Social Criticism* 49(10): 1244–1261.

Dabbagh, H., Mirdamadi, S. Y. and R. R. Ajani. 2023. Approaches to Muslim Biomedical Ethics: A Classification andCritique. *Bioethical Inquiry* 20: 327–339. https://doi.org/10.1007/s11673-023-10239-6

Darby, R. 2013. *A Surgical Temptation: The Demonization of the Foreskin and the Rise of Circumcision in Britain*. University of Chicago Press.

Darby, R. 2016. Moral Hypocrisy or Intellectual Inconsistency? A Historical Perspective on Our Habit of Placing Male and Female Genital Cutting in Separate Ethical Boxes. *Kennedy Institute of Ethics Journal* 26 (2): 155–163.

Dawson, A., A. Rashid, R. Shuib, K. Wickramage, M. Budiharsana, I. M. Hidayana, and G. Marranci. 2020. Addressing Female Genital Mutilation in the Asia Pacific: The Neglected Sustainable Development Target. *Australian and New Zealand Journal of Public Health* 44 (1): 8–10. https://doi.org/10.1111/1753-6405.12956.

Douglas, M., and A. Nyembezi. 2015. *Challenges Facing Traditional Male Circumcision in the Eastern Cape*. Human Sciences Research Council. Report available at http://www.hsrc.ac.za/uploads/pageContent/6391/.

Duivenbode, R., and A. I. Padela. 2019. The Problem of Female Genital Cutting: Bridging Secular and Islamic Bioethical Perspectives. *Perspectives in Biology and Medicine* 62 (2): 273–300.

Earp, B. D. 2016. Between Moral Relativism and Moral Hypocrisy: Reframing the Debate on "FGM". *Kennedy Institute of Ethics Journal* 26 (2): 105–144.

Earp, B. D. 2019. Mutilation or Enhancement? What is Morally at Stake in Body Alterations. Practical Ethics Blog: https://blog.practicalethics.ox.ac.uk/2019/12/mutilation-or-enhancement-what-is-morally-at-stake-in-body-alterations/

Earp, B. D. 2021. Male or Female Genital Cutting: Why 'Health Benefits' are Morally Irrelevant. *Journal of Medical Ethics* 47 (12/e92): 1–10.

Earp, B. D. 2022. Against Legalizing Female 'Circumcision' of Minors. *Global Discourse* 12 (1): 47–76.

Earp, B. D. 2022b. Genital Cutting as Gender Oppression: Time to Revisit the WHO Paradigm. *Frontiers in Human Dynamics* 4. https://doi.org/10.3389/fhumd.2022.778592.

Earp, B. D., and R. Darby. 2017. Circumcision, Sexual Experience, and Harm. *University of Pennsylvania Journal of International Law* 37 (2): 1–57.

Earp, B. D., and S. Johnsdotter. 2021. Current Critiques of the WHO Policy on Female Genital Mutilation. *International Journal of Impotence Research* 33:196–209.

Earp, B. D., J. Hendry, and M. Thomson. 2017. Reason and Paradox in Medical and Family Law: Shaping Children's Bodies. *Medical Law Review* 25 (4): 604–627.

Earp, B. D., A. Shahvisi, S. Reis-Dennis, and E. Reis. 2021. The Need for a Unified Ethical Stance on Child Genital Cutting. *Nursing Ethics* 28 (7–8): 1294–1305.

Fahmy, M. A. B. 2015. *Rare Congenital Genitourinary Anomalies*. Springer Berlin Heidelberg. https://doi.org/10.1007/978-3-662-43680-6.

Fahmy, M. A. B. 2020. *Normal and Abnormal Prepuce*. Springer International Publishing. https://doi.org/10.1007/978-3-030-37621-5.

Gaut, B. 2002. Justifying Moral Pluralism. In Ethical Intuitionism: Re-evaluations, edited by P. Stratton-Lake, 137–160. Oxford: Oxford University Press.

Ghassemi, G. 2009. Criminal Punishment in Islamic Societies: Empirical Study of Attitudes to Criminal Sentencing in Iran. *European Journal on Criminal Policy* 15:159–180.

Gillon, R. 2015. Ethics Needs Principles—Four Can Encompass the Rest—and Respect for Autonomy Should Be "First Among Equals". *Journal of Medical Ethics* 41 (5): 575–578.

Gleave, R. 2018. Imami Shi'I Legal Theory: From its Origins to the Early Twentieth Century. In *The Oxford Handbook of Islamic Law*, edited by Anver M. Emon and Rumee Ahmed. Oxford University Press.

Glick, L. B. 2005. *Marked in Your Flesh: Circumcision from Ancient Judea to Modern America*. Oxford: Oxford University Press.

Gollaher, D. L. 2000. *Circumcision: A History of the World's Most Controversial Surgery*. New York: Basic Books.

Gomaa, A. 2013. The Islamic View on Female Circumcision. *African Journal of Urology* 19 (3): 123–126.

Hallaq, W. 1997. *A History of Islamic Legal Theories: An Introduction to Sunni Usul al-fiqh*. Cambridge University Press.

Hodson, N., B. D. Earp, L. Townley, and S. Bewley. 2019. Defining and Regulating the Boundaries of Sex and Sexuality. *Medical Law Review* 27 (4): 541–552. https://doi.org/10.1093/medlaw/fwz034.

Johnson, M. 2010. Male Genital Mutilation: Beyond the Tolerable? *Ethnicities* 10 (2): 181–207.

Kassamali, N. 2005. Genital Cutting. In *Encyclopedia of Women and Islamic Cultures: Family, Body, Sexuality and Health*, edited by Suad Joseph and Afsaneh Najmabadi, vol. 3, 129–132. Brill.

Khomeini, R. 1987. *Tahrir Al-Vasilah*. Dar Al-Qalam Publisher.

Kigozi, G., M. Wawer, A. Ssettuba, J. Kagaayi, F. Nalugoda, S. Watya, F. W. Mangen, N. Kiwanuka, M. C. Bacon, T. Lutalo, D. Serwadda, and R. H. Gray. 2009. Foreskin Surface Area and HIV Acquisition in Rakai, Uganda (Size Matters). *AIDS (London, England)* 23 (16): 2209–2213. https://doi.org/10.1097/QAD.0b013e328330eda8.

Macklin, R. 2003. Applying the Four Principles. *Journal of Medical Ethics* 29 (5): 275–280.

Mernissi, F. 1992. *The Veil and the Male Elite: A Feminist Interpretation of Women's Rights In Islam*. Basic Books. Reprint edition.
Mulla Muhsin al-Fayḍ al-Kāshānī. 1960. *Al-Mahajjat al-Bayda fi Tahdib al-Ihya*. Tehran: Maktabat al-Sudouq.
Myers, A., and B. D. Earp. 2020. What is the Best Age to Circumcise? A Medical and Ethical Analysis. *Bioethics* 34: 645–663.
Najafi, M. H. 1981. *Javaher Al-Kalam fi Sharhe Sharaye Al-Islam*. Dar Ehya Al-Toras Al-Arabi Publisher.
Njambi, W. N. 2004. Dualisms and Female Bodies in Representations of African Female Circumcision: A Feminist Critique. *Feminist Theory* 5 (3): 281–303.
O'Connell, H. E., N. Eizenberg, M. Rahman, and J. Cleeve. 2008. The Anatomy of the Distal Vagina: Towards Unity. *The Journal of Sexual Medicine* 5 (8): 1883–1891. https://doi.org/10.1111/j.1743-6109.2008.00875.x.
O'Neill, S., D. Bader, C. Kraus, I. Godin, J. Abdulcadir, and S. Alexander. 2020. Rethinking the Anti-FGM Zero-Tolerance Policy: From Intellectual Concerns to Empirical Challenges. *Current Sexual Health Reports* 12 (1): 266–275. https://doi.org/10.1007/s11930-020-00299-9.
Oba, A. A. 2008. Female Circumcision as Female Genital Mutilation: Human Rights or Cultural Imperialism? *Global Jurist* 8 (3): 8.
Odukogbe, A. A., B. B. Afolabi, O. O. Bello, and A. S. Adeyanju. 2017. Female Genital Mutilation/Cutting in Africa. *Translational Andrology and Urology* 6 (2): 138–148.
Pippi Salle, J. L., L. P. Braga, N. Macedo, N. Rosito, and D. Bagli. 2007. Corporeal Sparing Dismembered Clitoroplasty: An Alternative Technique for Feminizing Genitoplasty. *The Journal of Urology* 178 (4 Suppl): 1796–1801.
Purpura, V., E. Bondioli, E. J. Cunningham, G. De Luca, D. Capirossi, E. Nigrisoli, T. Drozd, M. Serody, V. Aiello, and D. Melandri. 2018. The Development of a Decellularized Extracellular Matrix-Based Biomaterial Scaffold Derived from Human Foreskin for the Purpose of Foreskin Reconstruction In Circumcised Males. *Journal of Tissue Engineering* 9 (4): 2041731418812613.
Rashid, A., Y. Iguchi, and S. N. Afiqah. 2020. Medicalization of Female Genital Cutting in Malaysia: A Mixed Methods Study. *PLoS Medicine* 17 (10): e1003303.
Rezaei, M. 2020. A Non-Fundamentalist Return to Origin: The New Islamic Reformers' Methodology of (Re)interpretation. *Critical Research on Religion* 8 (1): 25–38.
Ridgeon, L. 2020. Ahmad Qabil, a Reason to Believe and the New Religious Thinking in Iran. *Middle Eastern Studies* 56 (1): 1–15.
Rogers, J. B. 2016. The First Case Addressing Female Genital Mutilation in Australia: Where is the Harm? *Alternative Law Journal* 41 (4): 235–238.
Sachedina, A. 2009. *Islamic Biomedical Ethics Principles and Application*. Oxford University Press.
Scanlon, T. 2002. Rawls on Justification. In *The Cambridge Companion to Rawls*, edited by S. Freeman, 139–167. Cambridge University Press.
Shahrur, M. 2009. In *The Qur'an, Morality and Critical Reason*, edited by Andreas Christmann. Brill.
Shahvisi, A., and B. D. Earp. 2019. The Law and Ethics of Female Genital Cutting. In *Female Genital Cosmetic Surgery: Solution to What Problem?* edited by S. Creighton and L.-M. Liao, 58–71. Cambridge: Cambridge University Press.
Shweder, R. 2022. The Prosecution of Dawoodi Bohra Women: Some Reasonable Doubts. *Global Discourse*. 12 (1): 9–27. https://doi.org/10.1332/204378921X16141809582432
Soroush, A. 2009. *The Expansion of Prophetic Experience*. Brill.
Sorrells, M. L., J. L. Snyder, M. D. Reiss, C. Eden, M. F. Milos, N. Wilcox, and R. S. Van Howe. 2007. Fine-Touch Pressure Thresholds in the Adult Penis. *BJU international* 99 (4): 864–869.
Syamsuddin, S. 2001. Abū Ḥanīfah's Use of the Solitary Ḥadīth as a Source of Islamic Law. *Islamic Studies* 40 (2): 257–272.

Taher, M. 2017. *Understanding Female Genital Cutting in the Dawoodi Bohra Community: An Exploratory Survey. Sahiyo:* https://www.fgmcri.org/academic_repository/understanding-female-genital-cutting-in-the-dawoodi-bohra-community-an-exploratory-survey/

Tangwa, G. B. 2004. Bioethics, Biotechnology and Culture: A Voice From the Margins. *Developing World Bioethics* 4 (2): 125–138.

ToubiaN. F. (1999). Evolutionary cultural ethics and the circumcision of children, in G. C. Denniston, F. M. Hodges, and M. F. Milos, eds., *Male and Female Circumcision* (Boston, MA: Springer), 1–7. https://doi.org/10.1007/978-0-585-39937-9_1

Tusi, J. 1973. *Al-Mabsut fi Figh Al-Imimya*. Al-Makteba Al-Mortazaviah Publisher.

van der Merwe, A. 2020. In Response to an Argument Against Penile Transplantation. *Journal of Medical Ethics* 46 (1): 63–64.

Werker, P. M. N., A. S. C. Terng, and M. Kon. 1998. The Prepuce Free Flap: Dissection Feasibility Study and Clinical Application of a Super-Thin New Flap. *Plastic and Reconstructive Surgery* 102 (4): 1075–1082.

WHO. 2008. Eliminating female genital mutilation: an interagency statement. World Health Organization: https://www.who.int/publications/i/item/9789241596442

Chapter 13
Enhancement, Natural Law, and New Basic Human Goods

James J. Delaney

Abstract This chapter explores the permissibility of radical human enhancements within the framework of Thomistic natural law, focusing on their potential to expand access to existing basic human goods and to introduce entirely new ones. Arguing against traditional concerns that enhancements might violate human nature or diminish human flourishing, the author contends that enhancements consistent with natural law principles, such as respecting incommensurable goods and avoiding the intentional destruction of existing goods, could uniquely enrich human life. Addressing metaphysical objections, the chapter maintains that enhanced capacities can coexist with human nature as rational animals. Finally, the Christian perspective is considered, examining how enhancements align with or challenge the ultimate end of union with God, concluding that cautious engagement rather than outright rejection is the appropriate response for Christian and other natural law theorists.

13.1 Introduction

In this chapter, I want to explore some questions about human enhancement from a natural law perspective. The sort of natural law theory I have in mind has its roots in the thought of Saint Thomas Aquinas. It is an ethical theory, although as I will argue, it is better understood more broadly as theory of practical reason. And like Aquinas I'll work from a hylomorphic understanding of the nature of human persons. On this view, going back to Aristotle, human persons are essentially form-matter composites; animals informed by a rational soul. The precepts of natural law/practical reason are derived from human nature itself. That is, what human beings have reason to pursue (as well as avoid) is determined by the sort of creatures that we are, and by that which constitutes our flourishing. And while it is certainly not the case that *all* Christians accept this Thomistic framework, I think its basic tenets

J. J. Delaney (✉)
Department of Philosophy, Niagara University, Lewiston, NY, USA
e-mail: jdelaney@niagara.edu

K. R. Smith (ed.), *Bioethics and Religion*, The International Library of Bioethics 116, https://doi.org/10.1007/978-3-032-33538-8_13

cohere with a Christian worldview. Furthermore, it forms the philosophical basis for a good deal of official Catholic moral and metaphysical doctrine. Therefore, my hope is that by using the lens of Thomistic natural law theory, I can offer something in the way of a Christian perspective. It should be noted that one need not be a Christian to accept the core tenets of natural law theory (at least the version of the so-called "new natural law theory" that will be my focus). Nevertheless many, and probably even the majority, of those who hold the view are also Christian. These core natural law tenets are not only compatible with, but indeed quite complementary to, some of the most foundational beliefs that Christians hold. I have three particular beliefs in mind. First, Christians believe that the ultimate end for human beings is eternal union with God. Second, this union is made possible by Christ's sacrifice and thus salvation comes through Christ. And finally, eternal salvation is not found in the earthly life.

Human enhancement is of course a much-discussed topic in bioethics. There are disputes about how exactly we ought to define it. But for my purposes here, I just mean the improvement or addition of various sorts of capacities that human beings possess. These improvements in turn contribute to the increased performance of various functions and activities. The functions could be physical (such as increased athletic ability, resistance to disease, or increased life expectancy), cognitive (such as increased intelligence, memory, or analytical skill), or moral (such as increased empathy or impulse control). Furthermore, the improvements are produced by various sorts of biotechnological interventions. These include things like genetic manipulation, various sorts of drugs, and nanotechnology. Finally, enhancements are typically distinguished from therapy, which involves the use of biotechnological interventions to treat disease or injury.

One class of enhancements are those that would raise the capacity of an individual beyond the standard or healthy range to a level that some other human beings currently possess. They would include, for example, raising someone's strength or athleticism to that of a world class athlete, or raising certain of one's cognitive abilities to the levels possessed by those at the highest end of the spectrum. A second class of enhancements, what I term "radical enhancements," are those that would raise capacities beyond those ever possessed by *any* human being. Consider athletic ability. Usain Bolt's world record in the 100-m dash is 9.58 s. A radical enhancement would increase capacities so that a person could run the 100 m in, say, 5 s. Increasing cognitive capacities beyond those ever possessed by any human is another example. Some proponents of radical enhancement imagine human beings with the ability to read a novel in a matter of minutes. Or consider life expectancy. Suppose we could use biotechnology so as to allow human beings to live to be a 1000 years old on average. This too would be a radical enhancement. In addition to improving our current capacities, radical enhancements could also involve creating new capacities altogether. The ability to communicate telepathically is an example.

To the extent that radical enhancement has been discussed among natural law theorists, the positions are mixed. Some of those sympathetic to natural law seem to think that radical enhancements constitute a violation of human nature. If this is true, then there is not much need for further discussion. For if we understand human nature in terms of the fundamental capacities that humans currently possess, and radical enhancements fundamentally change or add to those capacities, then radical enhancement damages rather than improves nature. By analogy, one might think that it would be misguided to try to "improve" a great novel or work of art. As Jensen and Widow put it recently, many enhancements produced by technological means such as genetic engineering or drugs constitute a "movement against nature" and should be rejected (2018).[1] Or consider the following statement from the Catholic Church's International Theological Commission with respect to using germ-line genetic engineering for enhancement purposes:

> The use of genetic modification to yield a superhuman or being with essentially new spiritual faculties is unthinkable, given that the spiritual life principle of man—forming the matter into the body of the human person—is not a product of human hands and is not subject to genetic engineering. The uniqueness of each human person, in part constituted by his biogenetic characteristics and developed through nurture and growth, belongs intrinsically to him and cannot be instrumentalized in order to improve some of these characteristics (Congregation for the Doctrine of the Faith 2004, 91).

Despite these sorts of arguments, I hope to show that some radical enhancements might actually be permissible on natural law grounds. In fact, since they could potentially contribute to human flourishing in a unique way, they might actually be encouraged. The chapter will proceed as follows. First, I give a broad sketch of the basic tenets of natural law theory. Provided that the means are licit, I defend the moral permissibility of pursuing radical enhancements with a special focus on those which would give access to distinctly new kinds of basic human goods. I then turn a potential metaphysical objection that this sort of radical enhancement would fundamentally change human nature. I offer a brief overview of the Thomistic hylomorphic account of human beings as animals informed by a rational soul. Against the objection, I claim that altering some of our capacities so as to gain access to new types of good is compatible with our remaining fundamentally the same sorts of creatures. Finally, I explore how some of the basic elements of Christianity, and specifically the foundational Christian beliefs I mention above, will affect how natural law theorists who are Christian understand the benefits and potential dangers of radical enhancement.

[1] It should be noted that Jensen and Widow are more concerned with the means involved in enhancement rather than the ultimate ends or dispositions that they intend to produce. However, in the case of many radical enhancements, it may be that the only way to achieve the ends will be by manipulating capacities that are unnatural on their account.

13.2 Natural Law, Basic Human Goods, and Radical Enhancement

Aquinas' theory of natural law begins with his distinction between speculative and practical reason. Speculative reason's first subject is "being" and concerns necessary subjects. Practical reason's first subject is "good" and concerns contingent matters such as human actions. Each type of reason is grounded in a "primary precept," which is self-evidently true. The primary precept of speculative reason is "The same thing cannot be both affirmed and denied at the same time." The primary precept of practical reason is that "Good is to be done and pursued, and evil is to be avoided" (Aquinas, 1947, ST I-II, q. 91 a. 2). There is debate among natural law theorists about the relationship between these two primary precepts. On the traditional interpretation, the primary precept of practical reason is thought to be derived from speculative knowledge. In this chapter, however, I will be working from the so-called "New Natural Law Theory" perspective.[2] On this view, the primary precept of practical reason is not derived, but rather is simply the grounding principle of agency (in the way that the principle of non-contradiction grounds speculative reasoning).[3] That is, deliberating about different courses of action that are open to an agent requires that the agent recognize that she ought to pursue that which is worth pursuing and avoid that which ought to be avoided. The question, then, is how one determines the sorts of things should be pursued and which should be avoided. Aquinas states:

> Since, however, good has the nature of an end, and evil, the nature of a contrary, hence it is that all those things to which man has a natural inclination, are naturally apprehended by reason as being good, and consequently as objects of pursuit, and their contraries as evil, and objects of avoidance. Wherefore according to the order of natural inclinations, is the order of the precepts of the natural law (Aquinas, 1947, ST I-II q. 94 a. 2).

Aquinas goes on to list several different things to which human beings are naturally inclined. Like all living things, there is a natural inclination to avoid death. In common with other animals, there is a natural inclination to reproduce and raise offspring. And as rational creatures, humans are inclined to know the truth about various subjects and live peaceably with one another. Contemporary natural law theorists typically approach the question of what Aquinas refers to as natural inclination in terms of *basic human goods*, propositions that are:

> …known in themselves and not through the mediation of some further proposition. Such propositions can be said to be self-evidently known, not because knowledge of them is innate, or because they are immediately known with certitude, but because reason, operating practically, recognizes the desirability, and hence actionability, the to-be-done-ness, of

[2] Beginning with Germain Grisez (1965, 1983, 1993), this theory is further articulated and developed in the work of Joseph Boyle (2002, 2005a, 2005b), John Finnis (1983, 1987, 2011), Patrick Lee (2019), Robert George (1999), and Christopher Tollefsen (2008, 2017) among others.

[3] There is additional debate between traditional and new natural law theorists regarding the ultimate end for human beings. I will say more about this in the final section of the chapter.

> the goods immediately upon a true apprehension of the nature of those goods (Tollefsen 2017, 135).

Basic human goods include items like life (and health), knowledge, friendship, excellence in work and play, practical reasonableness, and religion among others. There is some debate among natural law theorists regarding the precise number of items the list includes. However, there is general consensus about a many of them.[4] Before continuing, I should clarify the way in which natural law theorists conceive of two of the items on the list. It may seem strange for "practical reasonableness" to be listed among the goods, since natural law essentially just *is* a theory of practical reason. But the proper exercise of practical reason is itself a good, and thus something of value. Finnis calls it "the basic good of being able to bring one's own intelligence to bear effectively on the problems of choosing one's actions and life-style and shaping one's character…[It involves] freedom and reason, integrity and authenticity" (Finnis 2011, 88). In an important sense, the good of practical reasonableness can be understood as the good of autonomy; the ability to choose to live one's life in accordance with one's own self-chosen values.[5] The other item to clarify is "religion." As I stated above, many natural law theorists are Christian, but the good of religion is understood much more broadly. Murphy describes it as "being in harmony with the more than human order, those aspects of the way things are that transcend the world of human making, doing, and acting" (Murphy 2001, 131). Just as human beings are inclined by their nature to seek out friendships, excellences in their activities, etc., they are inclined to come to an understanding of the ultimate causes of themselves and their place the world. And, as Finnis states, "[it is] particularly important to have thought reasonably and (where possible) correctly about these questions of the origins of the cosmic order and of human freedom and reason—whatever the answer to those questions turns out to be, and even if the answers have to be agnostic or negative" (Finnis 2011, 89). The good of religion therefore is just the general sort of pursuit of answers to these deep questions, and is not constrained to any particular faith tradition.

Basic human goods are the most fundamental reasons for action. New natural law theorists hold that the basic human goods are *incommensurable*. There is no common measure to which they can be reduced and by which they can be compared against one another. There is also no hierarchy among the goods. That is, there is no sense in which one has reason to pursue one good over another because it is intrinsically superior. For example, suppose I am deliberating about whether to go to the gym (an instance of promoting health) versus spending the afternoon visiting with

[4] For a summary of the debates regarding competing catalogs of basic human goods, see Murphy 2019.

[5] Although as Jason Eberl points out, the natural law understanding of autonomy should not be understood in an individualistic liberal sense that would include even actions that were damaging to one's own well-being. Rather, autonomy is understood in a more Kantian sense: "…it is it is the will's capacity to self-legislate—that is, to govern oneself in accordance with the rationally understood moral law. Thus, an agent's exercise of her autonomy should lead to both individual and collective flourishing" (Eberl 2014, 292).

friends (an instance of promoting friendship). Either choice could be practically reasonable, although the unique circumstances might make it the case that I have more reason to choose one over the other. Perhaps I haven't been to the gym in a while or that I will not have another opportunity to visit with these friends again for a significant period of time. Regardless, there is no sense in which all other things being equal, I have *more* reason to pursue health than friendship. Finally, the incommensurability thesis holds that instances of the basic human goods are uniquely valuable. One cannot get that which is valuable about one good by possessing another. So even if I have ultimately have more reason to go to the gym than to visit with friends given the particulars of my situation, I still miss out on something of unique value (the value of *that specific instance* of promoting friendship). It should be noted that the same holds between different instances of the *same* good. Perhaps I am choosing between having coffee with one friend versus going on a hike with another. Both are instances of promoting friendship, but they too are uniquely valuable.[6] The incommensurable nature of the goods entails that there is no one way to lead a meaningful and flourishing life according to new natural law theory. People can (and do) prioritize goods in vastly different ways. In this sense, the theory is quite open. And the nature of the incommensurability points to an important point about the reality of agents' choices. Any choice that one makes to pursue a given instance of a good will necessarily rule out the choice to pursue innumerable instances of other goods instead. These other goods are, in a sense, knowingly ignored. But this in itself cannot make the choice practically unreasonable (for otherwise *every* choice would be unreasonable). The important thing is that "the practical principles direct one to love and respect the various human goods—as instantiated in oneself or in others—the moral principle is equivalent to the proposition that one ought morally to choose in line with love and respect for all the basic goods" (Lee 2019, 78). The principles of practical reasoning that govern the way in which basic goods may be pursued is what moves new natural law from a theory of action to a moral theory. Actions that presuppose something false about the nature of the goods are ruled out. The most well-known moral constraint is the prohibition of actions that have, as an ultimate or intermediate end, the intentional destruction of an instance of a basic human good. Murder and lying are standard examples. Such actions are morally impermissible regardless of the purported benefits they might produce, because they presuppose that the goods can be traded off against one another in the way that the natural law theorist denies is possible. This is why natural law places an emphasis on an agent's intention, for as we saw above, any choice to pursue some good inevitably leads to the failure to promote others. Therefore, natural law theorists typically accept some version of double-effect reasoning, which distinguishes between bad effects that are intended versus those that are not. Certain programs of radical enhancement would violate this principle. The most obvious sorts of examples would be things like killing innocent people in an effort

[6] Joseph Boyle argues that this sort of incommensurability is necessary in for free choice (2002, 2005a, 2005b).

to develop and perfect enhancement technologies. But we can imagine other cases. Suppose that we intentionally destroyed some people's capacity to realize an entire class of basic human goods in order that they could realize a larger stock of another class. Perhaps we eliminate or severely inhibit the capacity to develop friendship so that the affected people will devote themselves more fully to pursuing knowledge. These sorts of plans would clearly be ruled out on natural law grounds.

Intentional destruction is the most clear-cut way that an action can violate the principles of natural law. However, there are other more subtle ways that actions can be practically unreasonable. Murphy (2001) offers several. One rules out actions that are inefficient. Such actions use a greater amount of instrumental means to secure some good than some alternative that uses less. Some programs of radical enhancement could fall into this category. It may be that the goods a radical enhancement aims to capture could be more efficiently pursued by some other technological means rather than by increasing a human capacity.

A second principle prohibits actions that dismiss or devalue persons within the context of agent neutral ends. These are actions that treat a person's goods as less valuable than others in the context of agent-neutral reasons based on that person's identity. The principle demands that human beings are given equal respect. Any radical enhancement program would therefore need to show proper respect for all humans involved. This would involve justice concerns, so that some people would not be unfairly benefited over others.

Finally, actions that dismiss or devalue aspects of well-being are practical unreasonable. These would include plans that falsely presuppose that certain goods are not of unique value and worthy of pursuit. It can be difficult to know when the principle has been violated. Murphy suggests that it may be easier to recognize in the context of a global choice, a particular life plan, than in an individual choice. If an individual completed omitted one of the basic human goods from her life, and falsely believed that she was not missing out on anything of value, this would count as dismissing or devaluing an aspect of well-being. This illustrates a general aspect of natural law theory, one which Chappell (2004) argues distinguishes it most importantly from consequentialism: the significance of the attitude one takes toward the goods she is *not* pursuing. Regardless of the goods one chooses to pursue and prioritize, practical reason requires that the agent nevertheless have proper respect for the nature of *all* goods and an appreciation for their incommensurability and distinct values. Radical enhancement, I believe, could be in tension with this principle for reasons somewhat similar to those involved with the earlier discussion of intentional destruction. We could imagine a radical enhancement of some capacity (beyond what any human has ever possessed) that would dispose human beings toward one particular basic human good. Suppose, however, that the enhanced individuals became so disposed to pursuing instances of this good that they became overly dismissive of one or more of the others. Like the person in the example of the life plan, these individuals would presuppose something false about the nature of the goods. As noted above, it will be difficult to know when precisely a particular radical enhancement has violated the principle of devaluing/dismissing goods. But as a general guide, I think we ought to ensure that whatever enhancements we make, human beings have a reasonable appreciation of the various basic goods that they can pursue. In normal conditions, we should expect that they will choose to pursue and prioritize different basic human goods and their various instantiations with the same variety that we currently observe in (non-enhanced) human beings.[7]

The constraints have just outlined should certainly be taken seriously for the new natural law theorist. Nevertheless, I think it would be premature to conclude that all radical enhancements would violate the principles of practical reason. And it is important to note that the constraints are primarily concerned with the *means* employed. As we have seen, plans of action can be ruled out on natural law grounds even if the ultimate end they seek to produce is itself a good worth pursuing; one may not damage/devalue one good to produce another. But this raises a more fundamental question that I want to explore: could the goals of radical enhancement constitute such goods? I believe that the answer is yes, and that radical enhancement could positively affect basic human goods in distinct two ways.

The first involve giving human beings access to additional instances of basic human goods that would otherwise be unavailable to them. This could be direct. An enhanced capacity might enable a human being to access an athletic achievement or feat of knowledge that would simply be impossible with only our current capacities. It could also be indirect. That is, enhanced capacities could allow human beings to possess a larger *stock* of goods. For example, increased life expectancy could allow for a larger number of projects and relationships. Using enhancement technologies for the sake of gaining access to the currently existing items on the list of basic human goods certainly raises interesting questions. My own view is that, provided it was done within the constraints discussed above, radical enhancement of this sort is consistent with the natural law's principles of practical reason. But I won't argue for that conclusion here. Rather, my focus will be on the second way that I think radical enhancement could potentially contribute to human flourishing: giving humans access to instances of new basic human goods altogether.

What I mean by new human goods altogether is simply the expansion of the current catalog of goods. In addition to life and health, knowledge, friendship, etc., radical enhancement of the sort I am concerned with would give us new items, ones to which we are not currently inclined due to our limited capacities. These new goods would be of their own unique value, and incommensurable with the others. Would the addition of new basic goods make a positive contribution to human flourishing? Or, would this constitute the creation of a "new superhuman or being with essentially new spiritual faculties" we saw the Church condemn above as seriously immoral?

Admittedly the idea of gaining access to new basic human goods by way of radical enhancement is somewhat speculative. While I think it is fairly reasonable to believe that certain radical enhancements could allow humans to realize instances of basic current basic human goods that would otherwise be inaccessible, it is not as obvious that they make the realization of new basic human goods possible. Part of

[7] Elsewhere, I refer to this as a "perspective of balance."

the reason for this is, I think, the limits of what we can imagine. It's difficult to conceive of what a new basic human good would be. If one could, then presumably it would already be something that we would recognize as having reason to pursue. That is, it would *already* be an existing basic human good. Despite the fact that we cannot specify the basic human goods that we have not yet discovered, it's quite plausible to think that there could be such goods. At the very least, it seems premature to rule it out. As Chappell states:

> The list of basic human goods I offer is long and complex—but it certainly isn't complete. Indeed, it couldn't be completed, because of a remarkable fact about humans. This is that the horizons of what they are and what they do are open in a radical way. It's not merely possible for humans to discover (or create) new instances of basic human goods. Art is an example. There was, presumably, a time when there were humans but no art. Likewise justice or fairness. There was, presumably, a time when there were humans, but no such thing as (consciously adopted) justice between them. The list of basic goods available to humans is in fact impossible to close while humans are still developing as a species—as I take it they are (Chappell 1998, 44).

There is of course no guarantee that radical enhancement would lead to new types of basic human goods. But consider other goods that have been created or discovered, such as Chappell's example of art, which we might understand more generally as the basic good of appreciating beauty. They likely emerge due to environmental factors such as having more leisure time, and increased cognitive faculties that allowed humans to form new concepts and ways of experiencing the world around them. Many radical enhancements could lead to similar circumstances. And once a good is discovered, it becomes part of human flourishing: "once a human good, always a human good" (Chappell 1998, 44).

Is the potential for the realization of new forms of basic human goods a reason for pursuing radical enhancement? One might say no. Suppose there is some basic human good genuinely distinct from knowledge, friendship, excellence in work and play, etc. Call it "x." One could argue that since x is not a good to which we are naturally inclined, we have no reason to pursue access to it. Since it falls outside the fulfilment of our human capacities, it is not part of human flourishing. We would in a sense be using instrumental means that could instead be used to pursue actual goods to pursue something that is not itself something of value for humans, thus violating the inefficiency principle. The argument is plausible, but I think it fails.

Consider what it is for any item to be "good." Traditionally, natural law theorists have based their account on an Aristotelian notion of human flourishing. Basic human goods are good because they fulfill *human* capacities. But this raises a puzzle. Why is human flourishing itself a good thing? Chappell has argued that the Aristotelian account cannot give a satisfactory answer to this question, and the value is simply assumed. Thus, it is anthropocentric and dogmatic.

The problem with the Aristotelian view on the value of the goods is that it gets the order of explanation backward. Instead, we should take a more Platonic view of the goods. For example, knowledge is not a good *because* it fulfills human capacities. Rather, human flourishing is good (more valuable than say cockroach flourishing to use Chappell's example) because it involves knowledge:

> If you see things this way round—the Platonist way—then the question why human flourishing is such a good thing need not be left unanswered and unanswerable, as it seems to be on the anthropocentric approach. Beyond mere dogmatic assertion, there will be a reason that we can give as to why human flourishing is a good thing, in a way that cockroach or canine flourishing is not. The reason will be that human flourishing involves such great goods as knowledge and friendship, and cockroach and canine flourishing do not. The line of thought helps to point us in the direction of a better account of what it is for anything to be a good. The idea will be that the goods are not, or not essentially, aspects of human flourishing. What they are essentially is this: they are ideal standards that apply in the different normative dimensions (Chappell 2004, 107–8).

If we take this Platonic approach, then new goods such as x will be good for human beings not in virtue of them being constitutive of human flourishing, but rather good for *any* substance with the capacities to realize it. The fulfilment of human capacities will now include x and human flourishing will have simply expanded. And this is all the better for human flourishing itself.

If new goods were to be discovered or created, they would enrich human flourishing in a unique way. Imagine an alternative human history in which one of our current basic human goods was never discovered/created. What if humans had evolved in such a way that they never developed the capacity to experience beauty? They would still recognize the other goods as worthy of pursuit, and perhaps having one less good accessible would even allow them to devote more time to the remaining goods to a larger degree. Nevertheless, I think most would view it as unfortunate that humans did not gain access to the good of appreciating beauty. Their flourishing is not all that it could be. The same is true of a good like x. If it is possible that radical enhancement could allow access to x, then it is practically reasonable to favor it all other things being equal provided that it did not do so in a way that violated the principles discussed above (intentional destruction, inefficiency, etc.).

13.3 Human Nature and New Basic Human Goods

Still, one might wonder if increasing human capacities so as to access new basic human goods raises a metaphysical problem. Even if the addition new goods expands human flourishing in a positive way, perhaps the changes to the capacities that enhancement brings about to realize those goods fundamentally changes human nature. If so, then radical enhancement does not so much improve human flourishing but rather replaces human beings with new kind of substance with a different sort of nature. This might constitute the creation of a "new superhuman or being with essentially new spiritual faculties" referenced earlier that the Church condemns as seriously immoral. I think this is a real concern for many natural law theorists and Christians who accept its basic tenets. Nevertheless, I hope to show that at the very least it is not obvious that *all* radical enhancements giving access to new basic human goods would fundamentally change our nature.

Most natural law theorists, and those who favor a Thomistic philosophical worldview generally, subscribe to a hylomorphic view of human persons. Each human person is identical to a human being, which is defined as an individual substance of a rational nature. Following Aristotle, Aquinas understands substances as form-matter composites. The soul is the form of the body. And while human souls have intellectual capacities that are not reducible to material processes and are therefore subsistent, human beings (like all animals) have a material nature. This distinguishes the Thomistic view from a Cartesian substance dualism that simply identifies the person with an immaterial soul itself. Thus, human persons are essentially embodied creatures, specifically rational animals:

> …it is quite natural to think that we are human beings, animals of the species *homo sapiens*. Our animality is special, for we are rational animals, but we are animals all the same. This is not simply a bit of metaphysical speculation; it is true to how we experience ourselves, and how we experience our relations to other persons and to the world (Anderson and Tollefsen 2008).

Above, we saw that Aquinas identifies the various sorts of items to which humans are inclined with those shared by all living things, those shared by other animals, and those unique to human beings as rational creatures. These mark different ontological kinds:

> The ontological category rational animal…does not necessarily correspond exclusively with the biological category *Homo sapiens*. Biological species categories are subsets of the ontological categories rational animal, sentient animal, and living organism. For example, although dogs and cats are not members of the same biological species, they are members of the same ontological kind: sentient animal (Eberl 2012, 32).

Throughout the chapter, I have defined enhancement in terms of improving capacities. On the Thomistic account, human persons (and indeed all organisms) are, in virtue of their substantial form, possessed of various *natural capacities*:

> …a capacity may be construed as a "radical" or "natural" endowment that is constitutive of a being's essential nature…its presence—along with any other proper endowments—indicates the existence of a being with the relevant specific nature (Eberl 2020, 14).

The natural capacities that a substance has corresponds to these different ontological categories. On the hylomor phic account, natural capacities need not be immediately exercisable. They belong to a substance's nature in virtue of the kind of thing that it is. For example, a developing human being in utero has the natural capacity to see (which corresponds to its nature as a sentient animal), even though it must develop further in order to actualize it. As rational animals, human beings have intellectual capacities. These can be broadly categorized as the capacities for self-conscious thought and autonomous volition, essentially the exercise of speculative and practical reason discussed in the previous section. A change in a being's natural capacities *could* constitute a change in its ontological status:

> An optimal way to characterize a substantial form would be the essential configuration of the matter composing a material substance such that the substance bears a set of properties that define its essential nature as individuated by the form inhering in a particular quantity of the designated matter…As a result, a change of substantial form would entail a change in numerical identity—meaning that one thing ceases to exist and something new comes to exist composed of the same matter but having a distinct form—and perhaps also changing

its species membership—as opposed an accidental form such as "being red," which defines a way that a thing may or may not be without altering its essential nature or changing its numerical identity (Eberl 2020, 21).

What sort of change in capacities would result in a substantial change, i.e., a change that is identity-affecting? One, it seems to me, would be an intervention that fundamentally removes a given part of something's fundamental nature. Consider the possibility of digitally uploading one's consciousness, an intervention favored by some transhumanists. If this were possible, then the resulting intelligence would presumably exercise intellectual capacities. But it would no longer be an animal with the right sort of material nature.[8] Indeed, the Church seems to express just this concern about attempting to "abandon" the body:

> The right fully to dispose of the body would imply that the person may use the body as a means to an end he himself has chosen: i.e., that he may replace its parts, modify or terminate it. In other words, a person could determine the finality or teleological value of the body. A right to dispose of something extends only to objects with a merely instrumental value, but not to objects which are good in themselves, i.e., ends in themselves. The human person, being created in the image of God, is himself such a good (HIG, 82).

A substantial change could also take place by giving capacities to a substance that would place it in a different ontological category. Suppose we gave sensory capacities to a plant. It's true that the resulting substance would have additional capacities that were not present before. So, in this sense, its flourishing has expanded. But on the hylomorphic account, the nature would have changed. Despite the continuity of the matter involved, the original plant was simply replaced by a new substance with a new form. The same would be true if we attempted to give intellectual capacities to merely sentient animal. If a radical enhancement required the addition of capacities of a different ontological category to human beings, then similarly, we would be replacing ourselves with new beings with a different nature. And this I think would be troubling to natural law theorists and to many Christians generally. I will not argue for that thesis here. Instead, I will accept it as a line that ought not to be crossed, and ask a different question. Would all radical enhancements that resulted in the ability to access new basic human goods involve an ontological change in human nature? I think the answer is no, or at least it is not obvious that the answer yes.

If a radical enhancement is practically reasonable, then it must be properly oriented to basic human goods. When we consider the current list of goods, we immediately see that our current capacities correspond to the goods that we are able to realize. For example, consider our vegetative capacities for growth and nutrition. Like all living things, they are key to us realizing the goods of life and health. The particular capacities that different substances have will of course vary. Plants have the capacity to photosynthesize whereas humans do not. The same is true of different species of the same ontological kind. Some sentient animals have the sensory capacity for sight while others do not. Intuitively, changing a substance's capacities through a radical enhancement seems consistent with its remaining of the same

[8] Hershenov (2019) argues that this is why animalists should reject transhumanism.

nature. For example, suppose that we used biotechnology to change our digestive system so as to be able to process foods that would otherwise be toxic. We would have new, or perhaps increased capacities that no humans have every possessed. But it seems we would remain the same kind of rational animal as well as the same biological species. While more controversial, why could it not be the case that we remain rational animals (thus the same ontological category) but no longer members of the biological species homo sapiens? Provided these changes were made with proper respect for basic goods, it is not clear why they ought to be ruled out on natural law grounds.

Still, one might object that adding or improving upon capacities merely allows us to realize currently existing goods in a new way. The new digestive capacities just allows us to capture the same instances of health that the previous capacities did. The same might also be said about realizing new instances of currently existing goods (new instances of knowledge, excellence in work and play, etc.), which I discuss above. But, according to the objection, the *actual goods themselves* are determined by our ontological nature. And any intervention that resulted in capacities that allowed the realization of new basic human goods (such as good "x") would be akin to giving sensory capacities to a being with a vegetative nature. It would necessarily result in a substantial change.

The objection is plausible, but I do not believe it is decisive. It seems to rest on the idea that there is a kind of "one to one" relationship between capacities and goods. Furthermore, these distinct capacities are ontologically definitive in the sense that the possession or lack of one determines the nature of the substance. But none of this is consistent with the standard hylomorphic account. As we saw above, specific capacities vary by species, but are generally categorized as vegetative, sentient, and rational. Humans do not have a "knowledge capacity," a "friendship capacity," an "appreciation of beauty capacity," etc. Rather, the goods to which we are inclined involve the exercise of one or more of the general capacities we have. Knowledge involves a combination of rational and sentient capacities such as sensory functions. And exercising these capacities requires life and a degree of biological health. Improving upon capacities or adding new ones (whether they be vegetative, sentient, or rational) is still consistent with our remaining fundamentally the kind of creatures that we are: rational animals. And if such changes allowed us to access x, there is no reason in principle that a substantial change would necessarily take place.

13.4 Christianity and Concerns About Enhancement

I want to conclude with some comments about Christianity in the new natural law tradition and radical enhancement. My focus has been enhancements that would give access to new basic human goods altogether, but I think these thoughts are relevant to bio-enhancement generally. Even if everything I have argued is true, one might worry that I have focused only on the earthly life. But this fails to acknowledge the ultimate end for Christians, which is union with God. The goals of radical enhancement that I have described such as increased physical achievements, new forms of knowledge, and significantly increased health and life expectancy, etc. seem to be more in line with a secular worldview associated with advocates of transhumanism. But this is a quite different sort of response to the frailty of the human condition than the Christian view. As Waters states:

> Transhumanists and Christians agree, for instance, that the finite and mortal human condition is far from ideal. For transhumanists humans have fallen short of achieving their true potential, whereas for Christians humans have not yet become the kinds of creatures God intends them to be. In response both agree that humans require release from their current conditions. For transhumanists this release is attained through technological transformation, whereas for Christians humans are transformed by their life in Christ. Both agree that death is the final enemy, transhumanists conquer this foe by achieving immortality of endless time, whereas Christians are resurrected into eternity…[But] these similarities…are more apparent than substantive…Christian theology should greet transhumanism with at best skepticism and at worst grave caution (Waters 2011, 164–165).

Why "skepticism" and "grave caution?" Waters' concern is that Christians might believe that the lofty visions of bio-enhancement and transhumanism provide the means to the ultimate goal of human (or perhaps post-human) fulfilment. But for the Christian, those goals could never be the case. Ultimate fulfillment can only be achieved in eternal life with God. As Saint Augustine famously says in the Confessions, "Thou madest us for Thyself, and our heart is restless, until it repose in Thee" (Augustine, 2023*). In the Catholic tradition accepted by many new natural law theorists, this involves the heavenly resurrection of the body. Devoting ourselves toward new instances of basic human goods and realization of perhaps even many new goods altogether could potentially confuse or distract us from our true ultimate end. One may worry that religion itself (which is one of the goods) could potentially be relegated to a lesser priority just due to the fact that there would be so many new additional sorts of goods available to human beings.*

The first thing to be said about this objection is that critics see it as a problem with new natural law theory generally apart from any particular worries about human enhancement. It marks one of the most fundamental disagreements between "traditional" and "new" natural law theorists. On the traditional view, the ultimate end for human beings is God alone, which is attained by the beatific vision. Therefore, God and the good of religion would hold a superior sort of status among the rest of the basic human goods. But as we have seen, new natural law theorists hold that there is no hierarchy among the goods. While religion is among the goods and provides a practically reasonable end to pursue, it does not have this elevated status. New natural law theorists who are Christian have offered a number of responses to the traditional criticism. Most importantly, they reframe the idea of the ultimate end as something that includes the basic human goods to which we are inclined.[9] To deny this, they argue, would fail to acknowledge the kinds of creatures that we are:

> Our ultimate end is not, therefore, the beatific vision, but a state of affairs that includes all persons with whom or for whose sake we can act, including God, with whose creative activity we cooperate in pursuing basic goods. Grisez calls this state of affairs "integral human fulfillment." By revelation we can know that we are promised the immortality necessary to achieve this state of affairs, which adequate reflection reveals to be dynamic and increasing in perfection, rather than static and "complete,' or unimprovable (Tollefsen 2008, 16).

Strictly speaking, new natural law theorists who are Christian believe that the kingdom of heaven is a condition that is "partly supernatural and partly natural" (Lee 2019, 87). While they emphasize the basic human goods as constitutive of our natural flourishing, they also acknowledge the importance of divine grace and revelation as necessary in order for us to reach our true fulfillment. And while religion is not superior to the other goods, it nevertheless plays a unique role for Christians. Lee has termed this an "organizational role:"

> When one becomes aware and reflects on the fact that the basic moral norms have God as their source—that God is the creator and so the source both of the natural inclinations directing one to the basic human goods, and of one's intelligence, by which one understands those directives—then one understands that one's choosing fully in line with the basic practical principles is a way of cooperating with God, and that failing to do so is disobeying God...Thus, the basic good of religion has an organizational priority to the other basic goods, given that the good of religion is at stake in all of one's actions (something not true of the other goods). All that one does should be a way of cooperating with, rather than setting aside the wisdom of God the Creator (Lee 2019, 85).

Let us return to the objection that a program of radical enhancement could threaten humans' focus on their ultimate end. It seems clear that this should be a genuine concern for the Christian new natural law theorist. For even if enhancement contributes positively to human flourishing according to our nature, the pursuit of an ever-increasing number of new goods could indeed become a sort of distraction. One could easily imagine a world in which we became so consumed with improving the human condition that religion failed to play the crucial organizational role Lee describes. There is a related worry, most notably articulated by Michael Sandel (2007) about the sort of mentality that radical enhancement promotes. Nature itself becomes something merely to be mastered and manipulated. This threatens our sense of humility and our appreciation of various aspects of our lives as gifts, values that Christians generally share. These sorts of concerns should give Christians who are new natural law theorists significant pause when thinking about the potential benefits of radical enhancement.

Nevertheless, I think it would be a mistake to conclude that the worry over radical enhancement's effect on religion and Christian values rules it out in principle. Why think that, *necessarily*, a human life that included a larger stock of instances of current basic human goods and new goods altogether would have such negative

[9] For a comprehensive account of the new natural law understanding of the ultimate end for human beings, see Grisez (2008).

effects? For even if one did have access to all of these additional goods, they would, in virtue of being practical agents, still be faced with the same reality. Recall that according to new natural law theory there are a vast number of particular choices among goods. One must choose a particular life plan that involves various sorts of projects, commitments, and relationships. Given the incommensurability of the goods, "there are indefinitely many good kinds of person to be, and…it makes little or no sense to regard any of these kinds as unqualifiedly the best kind of person to be" (Chappell 1998, 77). Radical enhancement could expand the range of choices about the kind of life one wants to live, but for the Christian, the good of religion could still play its foundational role. It is still "at stake in all of one's actions" and it seems that radically enhanced humans could remain committed to the importance of faith, revelation, and (for Catholics) sacramentality. Most importantly they could continue to hold the belief that our true ultimate end is union with God. By analogy, consider the ways in which other technological advances have affected human lives. Modern medicine allows us to live longer and better combat disease. Advances in science have provided the opportunity to achieve new feats of knowledge that were previously inaccessible. And despite some of its negative effects, an online world allows us to form and maintain friendships on a level we have never been able to previously. As proponents of enhancement often point out, most people including Christians think these are benefits. Why should it matter if similar benefits come about as a result of bio-enhancement? Like these other technologies, I think the proper response to radical enhancement for Christian new natural law theorists is one of caution and humility, not a rejection altogether.

13.5 Conclusion

My goals in this chapter have been modest in a sense. I have tried to show that some forms of radical enhancement, particularly those that would give access to new types of basic human goods, could be permissible for those (including Christians) committed to the basic tenets of new natural law theory. But many questions remain unanswered. How can we know when a radical enhancement program runs afoul of the prohibition on dismissing/devaluing other basic human goods? How can we determine precisely whether changes in very specific capacities might affect ontological status? If we pursue radical enhancement, how can Christians be confident that they will remain committed to their central values and the idea of God as the ultimate form of human fulfillment? Giving definitive answers to these questions goes well beyond the scope of this chapter. But if I am right, then they warrant further discussion because simply dismissing radical enhancement outright would be premature.

References

Anderson, Ryan, and Christopher Tollefsen. 2008. Biotech Enhancement and Natural Law. *The New Atlantis* 20:79–103.

Aquinas, Thomas. 1947. *Summa Theologica. Fathers of the English Dominican Province Translation. New York, NY: Benzinger Brothers.*

Augustine. 2023. *Confessions. E.B. Pusey, Trans. Project Gutenberg*. https://www.gutenberg.org/files/3296/3296-h/3296-h.htm. Accessed 23 August 2023.

Boyle, Joseph. 2002. Free Choice, Incomparably Valuable Options, and Incommensurable Categories of Good. *American Journal of Jurisprudence* 47:123–142.

Boyle, Joseph. 2005a. Being Reasonable in Choosing Among Incommensurable Goods. *Vera Lex* 6:11–34.

Boyle, Joseph. 2005b. Free Choice, Incommensurable Goods and the Self-Refutation of Determinism. *American Journal of Jurisprudence* 50:139–164.

Chappell, Timothy. 1998. *Understanding Human Goods*. Edinburgh: Edinburgh University Press.

Chappell, Timothy. 2004. The Polymorphy of Practical Reason. In *Human Values: New Essays on Ethics and Natural Law*, edited by D. Oderberg and T. Chappell. New York: Palgrave Macmillan Press.

Congregation for Doctrine of the Faith. 2004. *Humans in the Image and Likeness of God.* https://www.vatican.va/roman_curia/congregations/cfaith/cti_documents/rc_con_cfaith_doc_20040723_communion-stewardship_en.html. Accessed August 23, 2023

Eberl, Jason. 2012. Ontological Kinds Versus Biological Species. *The American Journal of Bioethics* 12 (9): 32–34.

Eberl, Jason. 2014. A Thomistic Appraisal of Human Enhancement Technologies. *Theoretical Medicine and Bioethics* 35:289–310.

Eberl, Jason. 2020. *The Nature of Human Persons*. Notre Dame, IN: University of Notre Dame Press.

Finnis, John. 1983. *Fundamentals of Ethics*. Washington DC: Georgetown University Press.

Finnis, John. 2011. *Natural Law Natural Rights*. 2nd ed. S New York: Oxford University Press.

George, Robert. 1999. *In Defense of Natural Law*. New York: Oxford University Press.

Grisez, Germain. 1965. The First Principle of Practical Reason: A Commentary on the Summa Theologiae, 1–2, Question 94, Article 2. *Natural Law Forum* 10:168–201.

Grisez, Germain. 1983. *The Way of the Lord Jesus, Volume I: Christian Moral Principles*. Chicago: Franciscan Herald Press.

Grisez, Germain. 1993. *The Way of the Lord Jesus, Volume II: Living a Christian Life*. Chicago: Franciscan Herald Press.

Grisez, Germain. 2008. The True Ultimate End for Human Beings: The Kingdom, Not God Alone. *Theological Studies* 69:38–61.

Hershenov, David. 2019. Why Transhumanists Can't Survive the Death of Their Bodies. *Ethics, Medicine, and Public Health* 10:102–110.

Jensen, J. and Jose Widow. 2018. Unnatural Enhancements. *Irish Theological Quarterly* 83 (4): 347–364.

Lee, Patrick. 2019. The New Natural Law Theory. In *The Cambridge Companion to Natural Law Ethics*, edited by Tom Angier. Cambridge: Cambridge University Press.

Murphy, Mark. 2001. *Natural Law and Practical Ethics*. New York: Cambridge University.

Murphy, Mark. 2019. *The Natural Law Tradition in Ethics*. The Stanford Encyclopedia of Philosophy. https://plato.stanford.edu/entries/natural-law-ethics/#CatBasGoo. Accessed 23 August 2023.

Sandel, Michael. 2007. *The Case Against Perfection: Ethics in the Age of Genetic Engineering*. Cambridge: Harvard University Press.

Tollefsen, Christopher. 2008. The New Natural Law Theory. *Lyceum* 10 (1): 1–17.

Tollefsen, Christopher. 2017. Natural law, basic goods, and practical reason. *In: The Cambridge Companion to Natural Law and Jurisprudence, edited by G. Duke and R. George. Cambridge: Cambridge University Press*

Waters, B. 2011. Whose Salvation? Which Eschatology? Transhumanism and Christianity as Two Competing Salvific Religions. In *Transhumanism and Transcendence: Christian Hope in an Age of Technological Enhancement*, edited by R. C. Turner. Georgetown University Press.

Epilogue

This book set out to address a fundamental and provocative question: **Should religion be given a place at the table of bioethics?** The contributors, representing both secular and religious perspectives, have offered a set of unique insights through their expertise in bioethics. Some explored this question in general terms, while others focused on specific, high-profile issues such as conscientious objection, physician-assisted suicide, abortion, circumcision, and religious influences on healthcare practices.

In this epilogue, I will not summarize the individual contributions but instead reflect on overarching themes and draw some personal conclusions. In so doing, I readily acknowledge that as a secular bioethicist, my interpretations are inevitably shaped by that perspective.

Defining the Terms

When posing the question of religion's place in bioethics, I deliberately refrained from defining either "religion" or "bioethics." These concepts are inherently contested, and leaving them open allowed contributors the flexibility to interpret them as they saw fit. Unsurprisingly, many authors felt compelled to clarify their understanding of these terms.

Religion, in particular, resists easy definition. Broad conceptions might reduce religion to nothing more than ultimate concerns or metaphysical frameworks, or, as in pantheism, equate it with everything. Such definitions risk stripping the term—and the concept of God—of their distinctiveness, leaving them devoid of the sacred or divinely ordained qualities that make them special and significant to their believers. Religion, as commonly understood, entails belief in a supernatural being or force that not only created the universe but also establishes moral laws, often conveyed through sacred texts. Major world religions such as Christianity, Judaism, and Islam share these characteristics and continue to exert significant influence across large segments of the global population.

K. R. Smith (ed.), *Bioethics and Religion*, The International Library of Bioethics 116, https://doi.org/10.1007/978-3-032-33538-8

A critical feature of these religions, and one that distinguishes them from secular frameworks, is the reliance on faith: belief in the absence of empirical evidence. While faith may provide guidance and solace for adherents, its divergence from reason and evidence-based thinking poses challenges for its integration into ethics, particularly in a pluralistic and scientifically informed society.

Religion and the Public Square

One recurring theme throughout this book is the notion of the "public square" as the proper arena for bioethical debate. Ethical discourse in the public square must, by definition, be accessible to all, regardless of personal beliefs. This requires grounding arguments in shared principles of reason, logic, and evidence. Secular contributors, unsurprisingly, emphasized this point most strongly. Yet most of the religious contributors also argued that their perspectives could be presented in terms accessible to a pluralistic audience.

Here lies a critical tension: religious advocates, in seeking to engage meaningfully in the public square, often reframe their arguments in secular terms. This raises a fundamental question: Does this approach represent genuine engagement, or is it a rhetorical adaptation that—whether unintentionally or deliberately—obscures underlying commitments to faith-based doctrines? The risk of philosophical inconsistency arises when arguments rooted in supernatural premises, which are not accessible to reason, are presented as though they stem from purely reasoned foundations. Whether this reflects unconscious bias, a well-intentioned misstep, or deliberate misrepresentation, such contributions may inadvertently weaken their credibility and pose challenges to the honesty and integrity of bioethical debate.

The Table of Bioethics

The "table of bioethics" is itself a metaphor open to interpretation. It may signify the academic domain, where professional bioethicists debate issues in journals and books. It could also encompass practical settings, such as government committees, hospital ethics boards, and public discourse, where bioethical principles influence real-world decisions. In liberal democracies, bioethical debates necessarily include diverse voices, and no one can—or should—be excluded based on their worldview.

The real question is not whether religious voices are allowed at the table, but whether their arguments merit equal weight. I would argue that religious claims, where rooted in faith and supernatural authority, cannot be fully accepted by secular participants or, indeed, by anyone committed to public reason. To participate meaningfully in bioethics, religious arguments must meet the same standards of evidence and rationality as secular ones.

Secularism and Moral Progress

Bioethics, as both a theoretical discipline and a practical endeavour, profoundly influences our lives. Secular frameworks, emphasizing autonomy, justice, and well-being, offer a foundation for ethical progress in pluralistic societies. These principles tend to promote individual freedom in matters such as reproductive choice, end-of-life decisions, and the protection of bodily integrity.

In contrast, religiously inspired arguments often lead to illiberal outcomes, including restrictions on abortion, contraception, and euthanasia. While religious voices should never be censored or suppressed, their claims must be subjected to the same rigorous scrutiny as secular arguments. If they fail to meet the standards of reason and evidence, they should not be accorded undue respect or influence.

Concluding Thoughts

This book's contributors have provided a diverse and insightful exploration of religion's role in bioethics. As editor, I have found their perspectives fascinating and thought-provoking, even when they challenge my own views. Ultimately, however, I remain convinced that bioethics must remain grounded in reason, evidence, and shared principles of public discourse.

Religion may occupy a seat at the table of bioethics, but its arguments must earn their place through reasoned justification, not appeals to faith or divine authority. Only by adhering to this standard can bioethics fulfil its promise of advancing human well-being, autonomy, and justice, and achieving moral progress in an increasingly secular world.

Index

K. R. Smith (ed.), *Bioethics and Religion*, The International Library of Bioethics 116, https://doi.org/10.1007/978-3-032-33538-8

GPSR Compliance

The European Union's (EU) General Product Safety Regulation (GPSR) is a set of rules that requires consumer products to be safe and our obligations to ensure this.

If you have any concerns about our products, you can contact us on ProductSafety@springernature.com

In case Publisher is established outside the EU, the EU authorized representative is:

Springer Nature Customer Service Center GmbH
Europaplatz 3
69115 Heidelberg, Germany

Batch number: 10918189

Printed by Printforce, the Netherlands